The ORIGIN of LANGUAGE

How We Learned to Speak and Why

MADELEINE BEEKMAN

SIMON & SCHUSTER
New York   Amsterdam/Antwerp   London
Toronto   Sydney/Melbourne   New Delhi

Simon & Schuster
1230 Avenue of the Americas
New York, NY 10020

For more than 100 years, Simon & Schuster has championed authors and the stories they create. By respecting the copyright of an author's intellectual property, you enable Simon & Schuster and the author to continue publishing exceptional books for years to come. We thank you for supporting the author's copyright by purchasing an authorized edition of this book.

No amount of this book may be reproduced or stored in any format, nor may it be uploaded to any website, database, language-learning model, or other repository, retrieval, or artificial intelligence system without express permission. All rights reserved. Inquiries may be directed to Simon & Schuster, 1230 Avenue of the Americas, New York, NY 10020 or permissions@simonandschuster.com.

Copyright © 2025 by Madeleine Beekman

All rights reserved, including the right to reproduce this book or portions thereof in any form whatsoever. For information, address Simon & Schuster Subsidiary Rights Department, 1230 Avenue of the Americas, New York, NY 10020.

First Simon & Schuster hardcover edition August 2025

SIMON & SCHUSTER and colophon are registered trademarks of Simon & Schuster, LLC

Simon & Schuster strongly believes in freedom of expression and stands against censorship in all its forms. For more information, visit BooksBelong.com.

For information about special discounts for bulk purchases, please contact Simon & Schuster Special Sales at 1-866-506-1949 or business@simonandschuster.com.

The Simon & Schuster Speakers Bureau can bring authors to your live event. For more information or to book an event, contact the Simon & Schuster Speakers Bureau at 1-866-248-3049 or visit our website at www.simonspeakers.com.

Interior design by Lewelin Polanco

Manufactured in the United States of America

1 3 5 7 9 10 8 6 4 2

Library of Congress Cataloging-in-Publication Data has been applied for.

ISBN 978-1-6680-6605-8
ISBN 978-1-6680-6607-2 (ebook)

*To my talkative daughters,
Janneke and Willemijn*

CONTENTS

INTRODUCTION: See How It Begins   1

PART ONE: MISTAKES WERE MADE . . .

ONE: The 1 Percent   13

    Same, Same but Different • Our Place in Nature • In Search of the Missing Link • *Pithecanthropus erectus* Becomes *Homo erectus* • Dwarf Sister or Distant Cousin? • A Revolution in Our Evolution • More Than Genes

TWO: Our Original Childcare Problem   45

    The Malleable Pelvis • When and Why Our Brain Became Big • On Babies • Hips Before Brains

THREE: Beat of a Different Drum   71

    Influential Families • Small Change, Big Effect • From 24 to 23 • The Beginning of Us

FOUR: Mind Blown   101

    An Heir and a Spare • Mind Expanding • The Importance of Staying Coolheaded • Molding a Head • An Evolutionary Spatial Packing Problem

PART TWO: ... AND THEN WE STARTED TALKING

FIVE: Loud Moms — 131

    A Brainy Problem • Self-Domestication • Who Cares? • All in the Family • Survival of the Cutest • Breaking Through the Gray Ceiling

SIX: Who Needs Half a Grammar? — 161

    The Language Gene • Who Needs a Miracle? • The Miracle Ape • Eye Power • Language as a Virus • Language Sculpts the Brain • A Little (Adaptive) Leg Up

SEVEN: Other Minds — 195

    Then There Was Language • The Ape That Gossiped • When *Was* There Language? • A Beautiful Mind Versus the Caring Mind • Back to My Imaginary Train Journey

EIGHT: Brave New World — 225

    The Moral Ape • Machines Like Us • From Caveman to the Moon in No Time • Man Is But a Worm • Curiosity Killed the Cat. Or Did It? • Ancient Information • A Unique Responsibility

EPILOGUE: Modern Family — 257

Acknowledgments — 263
Approximate Timing of Key Changes That
   Led Us to Become What We Are Today — 268
Notes — 271
Index — 297

The
ORIGIN
of
LANGUAGE

Introduction
SEE HOW IT BEGINS

At the time of her birth, my daughter weighed 5 pounds 13 ounces, around 5 percent of my own body mass. That doesn't sound like a lot, but compared to other mammals such an investment in a single baby is enormous. Yet, despite the significant resources my body allocated to her during the nine-month pregnancy, my daughter was born totally helpless. A typical human baby.

There was one thing she was particularly good at, though. Exploring my face. I distinctly remember spending ridiculous lengths of time simply staring into my baby daughter's little blue eyes, and she into mine. It felt as if we shared a secret language, searching each other's faces to get a sense of each other's feelings. Who are you? What are you thinking?

It turns out that my intuition was correct. Because our babies are so helpless when they are born, they need to be cradled, which brings their eyes very close to the face of whoever is holding them. And that provides the perfect vantage point from which to explore facial expressions of caretakers. So it is that little babies, my daughter included, start their journey to become masterful mind readers. By the age of one year, human babies have developed a theory of mind—the ability to understand what another person is thinking. They may not be able to walk unassisted, if at all, nor are they able to talk, but they sure know how to wrap others around their little finger.

Eighteen months after the birth of my first daughter, another girl was born. She was slightly heavier than her sister, but equally helpless and equally inquisitive. Now there were two little human beings totally dependent on us. Without knowing it, we, their parents, had embarked on a well-trod human journey: producing, often in quick succession, helpless babies with an innate ability to rapidly learn how to navigate the world around them. Helpless as they may be physically, our babies have exceptional emotional and psychological skills. They need those skills so they can build and foster strong social bonds with others around them. For babies, it is a matter of survival; they have to ensure the care they need for almost two decades. They don't just need their parents. Human babies are way too much work for two parents, let alone for one parent.

How did we get ourselves into such a precarious situation? This book is an investigation into how our babies became so helpless and the wonderful solution Mother Nature provided to help us cope. That gift from Mother Nature sounds as simple and ordinary as could be—effective communication—but how we came by it has puzzled many great minds for centuries.

Strange as it may seem, for me, this detective story started with ants.

◆

As I type these words at my desk at home, in the tropical north of Australia, ants wander around me in search of something to eat. At first their movements can look random, but when you focus on an individual ant, you soon start to see patterns in its behavior. Seeking food, it rarely walks in straight lines, instead slightly zigzagging over my desk, books, and laptop. The ants remind me of the Black Riders in Tolkien's *Lord of the Rings*. In search of Frodo, the Riders would stop regularly to sniff out the presence of Hobbits. Sitting high on their horses, they would slowly move their heads from left to right, inhaling deeply. The Riders' vision is poor, their olfactory sense acute. So it is with ants. Their world is one of smells. Smells guide them to potential food sources, but smells are also their main mode of communication.

The ant on my desk has just discovered a crumb of apple cake I carelessly spilled and couldn't be bothered to clean up. She first tries to see if she can carry it home herself, but soon realizes some help is needed. (I use "she" because all the ants you normally see are female.) Now, she will return to her nest in a razor-straight line, as there is no time to waste. She has to tell her nest mates to come help her before ants of a different colony discover the same treasure. Marking my desk with pheromones as she walks home, the ant ensures she'll be able to return quickly to the crumb. Soon more ants make their appearance at my leftover treat, each taking a bit of the crumb to carry back to the nest. Now there is so much ant traffic that a two-lane highway has formed, snaking over my desk, past my empty coffee cup and toward the cake. Each time two ants meet head to head, they stop briefly to touch each other with their antennae—which are basically the ants' noses—to confirm their identity. Do you smell like I do? Then fine, and we continue.

I've made a career of studying social insects—in particular, the way they "talk" to each other. Living in a society requires some form of communication, and the larger the society, the more complex its mode of communication. The ants feeding off my crumbs live in large colonies and construct trail networks using volatile pheromones that dissipate as soon as the ants stop reinforcing them. Because the ants live off ephemeral food sources, like my cake crumb, their trails disappear as soon as the food runs out. Otherwise, the tiny creatures would get "stuck" in their trail, wasting time walking to and from the now-depleted food source.

Ants that live in smaller colonies directly "tell" each other where the food is. One ant stumbles onto something to eat and then teaches her sisters where it is by slowly guiding them to the treat. She has to be slow about it, as her sisters will need to remember the route so that they, too, can teach nest mates. Others still—those that live in really tiny colonies—don't really talk to each other, with each ant doing her own thing. That works for them because there are so few of them.

Honeybees have more artistic style. They "talk" through dance. Austrian biologist Karl von Frisch shared the 1973 Nobel Prize in Physiology and Medicine for his deciphering of the symbolic "waggle dance" bees use to convey information about the direction and distance to a desired target. Dressed in traditional lederhosen, von Frisch would observe the individual bees he painstakingly marked with various combinations of paint colors. These days we use little discs with numbers that we glue onto the bees, but we still need the same patience von Frisch had all those years ago.

The honeybees von Frisch studied, the Western honeybee (*Apis mellifera*), is only one of the eleven or twelve species of honeybee (the precise number depends on whom you ask). This species has been kept by humans for thousands of years, and it so effectively spread across the globe that nowadays only Antarctica is free of the bee. All other species are found only in Asia and are mostly ignored, which is a great shame, as they are not only beautiful (I know I am biased) but also provide us with a window through which we can look at the evolution of bee communication. Together with my husband, who is as obsessed with bees as I am, and a bunch of students, I spent many months gluing numbers onto bees, or painting them when the bees were too small for our numbered discs, so that we could study them individually as they did their dance.

We wanted to learn how the precision of the bees' language depended on the difficulty of the problem they needed to solve. One might wonder why anyone would care about how well bees dance, but in a way their dance is equivalent to speaking and writing clearly. When it concerns humans, we know that precise communication is of the essence. For bees, precision matters too. Honeybees often go through a phase of homelessness as part of their reproductive cycle, or when they're following flowers in the tropics. Some species have specific requirements for their temporary housing: not too warm, not too cold, and most certainly dry. Only a hollow in a tree—or a suitable chimney, as a student of von Frisch discovered—will do. Other species are more flexible, happy enough with a twig, in the shade and away from ants. Others still, the giants of the honeybee world, need a sturdy branch or other strong structure capable of holding the significant weight of their substantial colony.

Finding a new suitable nest site and making sure your fellow bees are willing to relocate there presents a much more difficult problem than just settling anywhere for the time being. And as it turns out, bee species that never encounter a hard coordination problem can get away with a rather sloppy kind of waggling, whereas the picky species have to fine-tune their dance language. These results comport with my earlier research on ants. If nuanced communication is important, natural selection will refine the channels. After all, if you can get away with a bit of babble, why bother constructing grammatically correct sentences? Or the insect version of complicated speech.

My work as an evolutionary biologist has taught me to see all life forms equally, in the sense that everything alive is subject to natural selection and has evolved from a common ancestor. Humans included. An epiphany about the role of genetics in explaining the surprising biology of the South African Cape honeybee led to this book.

Without a doubt, the best place to study the Cape honeybee is the beautiful Cape Dutch village of Stellenbosch, located in the southernmost part of South Africa. Stellenbosch is known for its wonderful wine, making it a pleasant place to spend one's summers, sampling wines after a hard day's work. For over twenty years, my husband and I have been trying to understand the peculiar biology of the Cape bees. Their workers can create daughters by cloning themselves—a curiously self-involved development for a social insect, diverting as it does the workers' attention away from the interests of queen and colony, thereby changing, for the worse, the whole dynamic of the bee colony. After many years of experimenting, failing, and trying again, and drowning our sorrows with the help of Stellenbosch's wine and the unwavering good cheer of our South African colleagues, we found that everything can be explained by one small genetic change. That small genetic change, unique to the Cape bee, has huge flow-on effects on their behavior, turning the bees from (more or less) harmonious social critters into parasites of other bees, with devastating consequences for the South African beekeeping industry. Sometimes a seemingly tiny change can have an enormous effect. And, as I came to discover, as with bees, so with us.

◆

In essence, the human origin story is simple. More than 4 million years ago, our distant ancestors permanently swapped a life in the trees for one on two legs. Their skeletons changed, to make walking much easier and more efficient. Indeed, walking upright was a great innovation. It allowed our ancestors to literally walk out of their ancestral environment and start to explore the world. But nothing ever comes for free, and so the adaptations that made for efficient walkers also led to a series of problems—problems best addressed by forming strong social bonds. As with ants and bees, a social life requires some form of communication. Nothing too complicated initially—just enough to keep the group together. For a few million years, our human family expanded and spread around the globe. I refer here to *all* our ancestors, cousins and the like, to include every species of human-like creature that evolved after we split from the common ancestor we share with chimpanzees and bonobos. As soon as our species made its appearance, about 300,000 years ago, everything started to change. In no time, we had become the sole survivor; all others in our extended family swiftly went extinct. We have been exceptionally lucky. Truth is, we shouldn't be here.

Our species, *Homo sapiens*, has a design fault that should have been the end of us. As our skeletons adjusted to walking upright, our hips became narrower. Then, by a fluke of nature, our brain started to expand, forcing our head to change to make space for all that gray matter. And while changing our head was relatively easy—our skeleton had already shown how bony structures can change quickly—the consequences were significant. Babies with a large head and mothers with narrow hips do not make a good combination. But there is more. Growing a large brain is expensive—so expensive that mothers cannot allow their babies to stay in the womb until that brain is more fully grown. The consequence of this is that our babies are born early; their heads are too large, their brains are too expensive, and their mothers' birth canals are too narrow for them to be born any later than they are. Thus, the babies are seriously "underbaked." So underbaked, in fact, it takes them almost two decades to catch up, a time during which they have to rely on others.

So far, the story is not that new. Many have told similar origin stories, but I remained puzzled. Why did natural selection allow our species to get away with producing these useless and needy babies? I soon realized that the design fault that started the problem in the first place also provided its solution. That solution is language.

As the brain expanded and the head changed shape, the throat did too. The larynx descended, pulling the tongue with it. And that change led to two uniquely human characteristics. One is a high risk of choking on our food and drink, and the other is the ability to mold sounds like no other creature. Sounds we crafted into the language needed to organize our childcare.

I will argue here that talking and caring for underbaked newborns co-evolved, in an episode of runaway selection initiated by the genetic anomalies we now know spurred our species' neurological growth. Because here is another surprise. We are the result of a short series of discrete mistakes that made us what we are today. Our descent from other apes in the last few million years or so boils down to a mere handful of genetic and morphological turning points. These small changes incrementally built on each other, as in a multiplayer game in which the outcome is never quite predictable, setting in motion our transformation from tree-swinging apes to chatty apes.

The picture, in the end, is of an odd species that stumbled into global dominance through a relatively quick succession of simple mistakes. A species forced to speak up, or let its helpless infants perish.

The Origin of Language tells the story of how we learned to speak and why. It exposes a strong force—natural selection, solving problems by using whatever's lying around. A strong force that rarely over-delivers. If a simple solution will do the trick, natural selection usually won't bother with bells and whistles. And so we hone in on the small changes that made a huge difference in our evolutionary history.

Part I of the book uncovers how each fluke of nature became one step toward the creation of a new species. None of those flukes in and of themselves was sufficient, but slowly, over a period spanning 6 to 7 million years, cumulatively they led to a species with a unique ability for precise communication. This first part delves into the molecular roots of our unique talent for language; to dig into this seemingly miraculous accomplishment, we must use the sophisticated technical language of genetics. It can be a challenge, but I assure you it is worth it. Part II of the book then illustrates how this art of language kicked off an unstoppable process that led, ultimately and for better or for worse, to the global dominance of *Homo sapiens*.

I should emphasize that this is not a report of scientific consensus. We have no consensus, and perhaps never will, because the soft tissue involved in thinking and talking doesn't fossilize. And because it's often difficult to test whether a given feature was selected for or was just a "spandrel"—a side effect of something useful. But when you arrange the current puzzle pieces side by side, and apply Occam's razor, you find a new origin story for our species that was hiding in plain sight.

Or, perhaps, you'll see it in the eyes of the tiny, helpless, inquisitive baby you may be holding in your arms.

PART ONE
Mistakes Were Made . . .

one

THE 1 PERCENT

Nobel Laureate Niko Tinbergen once took a sabbatical from Oxford University and asked a young lecturer in biology to cover his teaching load. In an effort to connect with college students and animate the evolutionary relationship between organisms and genetic information, this resourceful stand-in started "thinking like a gene." What could you do, stuck on your home chromosome, to boost the chances you'd make it through to the next generation? So it was, from the lectern in 1966, that a young Richard Dawkins first pitched the gene's-eye view of evolution that would ultimately define his career as a public intellectual. By playing with alternative "units of selection," Dawkins was able to explain why it can appear that organisms are meticulously designed, even though there is no such thing as a designer.

Take a honeybee worker. When she stings a mammal in defense of her colony, she sentences herself to death, but ensures the propagation of the genes she shared with her sisters. Ever the genius, Charles Darwin had already speculated that *families* were biologically special when he considered the evolution of sterile insect workers. Self-sacrificial behavior, like dying in defense or refraining from reproducing, makes sense only if the beneficiaries of the action are family.

While Dawkins came to his insights during his stint as a stand-in lecturer, Darwin came to his while trying to avoid seasickness. Darwin spent five years sailing the world as "gentleman naturalist" on the HMS *Beagle*; his only official job was keeping the young captain Robert FitzRoy company during dinner. The *Beagle*'s previous captain, Pringle Stokes, had killed himself during a bout of depression in the waters off Tierra del Fuego, so the ship's owners thought it prudent to ensure FitzRoy had someone of his own class to converse with. (FitzRoy did ultimately kill himself, but only years after Darwin's trip.) Darwin took every opportunity to leave the ship and collect specimens that he later described, preserved, and took home to England. His extensive observations made him see that, contrary to religious doctrine, species were not immutable but, rather, changed in response to different environmental conditions. It was the beginning of what would become the most powerful idea within the biological sciences of the last two hundred years: evolution through natural selection.

Without knowing anything about genes (the word was "coined" only in the early twentieth century, years after Darwin had died), Darwin realized that if individuals of the same species vary in their reproductive success, he could explain how new species come about. That is, new species form from the accumulation, over thousands of generations, of small variations. And each small variation results in its bearer being just that little bit different than its ancestors, until ancestor and descendent are no longer alike. He may not have been totally correct, particularly when it comes to the accumulation of many tiny changes (as we'll see), but evolution by natural selection certainly explains the remarkable adaptations found in organisms.

To illustrate the power of evolutionary thinking, when Darwin received a specimen of a particularly weird-looking orchid, he predicted what sort of physical and behavioral characteristics its then unknown pollinator should have. That is, the pollinator should be able to reach down into the orchid's exceptionally long nectar spur. The actual pollinator was identified years after Darwin's death, and it had the exact characteristics predicted by him: a moth with an enormously long tongue. Evolution by natural selection also explains why bacteria become resistant to antibiotics, why head lice seem unfazed by insecticides, and why yeast is a good model system to investigate the growth of tumors.

Dawkins gave us a different view of evolution by shifting the focus from the organism, which was Darwin's focus, to its coalition of genes. His genius was his ability to picture the gene as the master of the universe. Genes best able to contribute to an individual that is well suited to survive and reproduce will win the evolutionary lottery. All that these genes "want" is to get into the next generation. In Dawkins's world, the organism is nothing more than a means to an end, a puppet to its masters. Dawkins's gene's-eye view did not challenge Darwin's theory; on the contrary, his view of genes as immortal keepers of information provided a mechanistic explanation for how natural selection could affect evolutionary change. It is perhaps no wonder, then, that I consider both Darwin and Dawkins my heroes, along with their respective books. Darwin's *On the Origin of Species by Means of Natural Selection, or the Preservation of Favoured Races in the Struggle for Life*, was first published in 1859; Dawkins's *The Selfish Gene* was published in 1976. I view both as evolutionary bibles, if that would not be a contradiction in terms.

Then came a revolution, the genomic revolution, a revolution that kicked off with the completion of the Human Genome Project in 2003. We can read a human genome! According to the Human Genome Project's website, the project gave us "the ability, for the first time, to read nature's complete genetic blueprint for building a human being." But did it? Not really. But that hasn't stopped some scientists from claiming otherwise.

The behavioral geneticist Robert Plomin is such a scientist. In his 2018 book *Blueprint: How DNA Makes Us Who We Are*, Plomin promises us a "fortune-telling device," a method that pinpoints exactly what genetic differences are responsible for which particular human traits. Alas, reading our DNA turns out to be much easier than figuring out what the collection of letters mean. The four letters of DNA's alphabet stand for the first letters of the four nucleotides, or bases—adenine, cytosine, thymine, and guanine—that make up DNA. The famous double helix of DNA forms because an A always forms a pair with a T, and a G always forms a pair with a C. The base pairs form the "ladder" structure of DNA. The human DNA ladder contains 3 billion steps. Sequencing a genome, then, involves no more than figuring out the order of the nucleotides. Sounds easy, and these days it is even cheap to sequence a genome. But what that order means is a different matter altogether.

With the exception of identical twins, we all have a unique DNA sequence. The smallest possible difference is a *single nucleotide polymorphism* (SNP, pronounced "snip")—a difference in a single "letter," or base pair. Imagine two fragments from different individuals, one reads AAGCCTA and the other reads AAGCTTA—a single difference in nucleotides, or C versus T. Because the nucleotides always form pairs, we know that the individuals also differ in the nucleotide found on the other DNA strand. Genome-wide association studies (GWAS) look for such minuscule differences between people in the hope of predicting the chance someone with a particular DNA profile will develop a disease such as diabetes or coronary heart disease. One of the first large-scale GWAS was published in 2007 by the UK-based Wellcome Trust Case Control Consortium. This study looked at a total of seven human diseases: coronary heart disease, type I diabetes, type II diabetes, rheumatoid arthritis, Crohn's disease, bipolar disorder, and hypertension.

A typical GWAS goes like this: Find a large group of people who suffer from a particular disease or disorder, and compare their genomes with the genomes found in a control group of people who are as similar as possible to the first group, but without the disease or disorder. Then trawl through 3 billion base pairs, the 3 billion steps on the human genome ladder, and find every SNP that differs in frequency between the two groups. Once found, apply a toolbox of statistical wizardry to produce a number that says something about the strength of the association between the genetic profile of the case group and the disease or disorder. The more individuals in the case group who share the same SNP that is absent in the control group, the stronger the association between the SNP and the disease. Then, combine all the SNPs that seem to be more frequent in the case group, use more statistical wizardry, and come up with a polygenic score—a number that reflects the association of many SNPs with a particular trait. Plomin's fortune-telling device relies on this exact logic, and in his own words, it "can tell your fortune from the moment of your birth, is completely reliable and unbiased—and costs only £100."

Alas, I would not recommend spending 100 pounds on a fortune-telling device, not because I do not believe that GWAS are unable to find associations between a particular genetic profile and whatever trait of interest, but because an association does not mean causation. One can draw an almost perfect correlation between the sales of organically produced food and the occurrence of autism, but no one in their right mind will argue that consumption of organic food causes autism. But there is more. Imagine you had your genome sequenced, and you were told you have a series of SNPs associated with developing type II diabetes. For the sake of argument, we assume the association between the genetic profile and disease is causal. You are given a number: your profile makes you 1.167 times more likely to develop type II diabetes than someone who does not have that particular combination of SNPs. The trouble is that such information is no more useful to an individual than is knowing you have a certain probability of dying in a car crash every time you drive your car. Calculating your chance of having a fatal car crash based on a large amount of data (statistics on the number of cars on the road, people per car, trips made, duration of trips, fatality rates, and so on) does not take into account your specific circumstances (sticking to the speed limit, only driving during off-peak traffic times, your age—you name it). Similarly, your type II diabetes SNP profile ignores the rest of your genetic makeup, your lifestyle, and your general health profile. There is a reason one of the most frequently used phrases in science is "all else being equal." All else is hardly ever equal.

Even so. Thanks to the many offshoots of the Human Genome Project, anyone who wants to and knows where to look, can now find huge amounts of genetic data on the internet. It would not surprise me if the study I am about to describe started as a joke, as the question it seems to ask is: Is there a genetic marker for income? Using data collected as part of the UK Biobank data set, this particular study found an association between 30 so-called loci (singular is *locus*, a particular location in the genome) and income in the UK. Well, that seems interesting, so the researchers dug deeper into the available data and identified another 120 loci associated with income. Given the looking-for-meaning-in-DNA frenzy, the researchers could see what sort of associations other studies had made regarding the newly identified "income loci." Lo and behold, 18 of the loci found to have an association with income had previously been associated with intelligence in other studies. Thus, if you have the right DNA, you are intelligent enough to achieve a high income. Right? Of course not, as the authors of the paper explain, but they do conclude that there are some genetic effects contributing to socioeconomic status in the UK. That last qualifier is important, as GWAS can only say something about the particular group of people, or population, under study. GWAS results do not generalize because there is so much more than just genes.

Does that mean we cannot derive anything sensible from looking at someone's genes? There are certainly some gems to be found in our genomes. Sometimes a single change can have huge effects. One such change determines whether your urine has a peculiar smell after a meal of asparagus. Surely, that's information you had been waiting for your whole life. (The culprit is methanethiol, a metabolite of asparagus. If you can't smell it, then you might be lucky, as it smells like rotten or boiling cabbage.) But the Human Genome Project did not turn out to be a Rosetta Stone for translating between genotype and phenotype. There will always be something lost in translation.

One big surprise from the Human Genome Project was the relatively small number of human genes. Before the project was finished, popular estimates had put the figure at roughly 100,000. The truth turned out to be less than 25,000, a fourth of that guess. To put these numbers into perspective, the Australian platypus has about 18,500 genes. So, we've done a lot with a little, especially considering how genetically similar we are to other species and to each other. Humans share about 98 percent of their DNA with pigs, 85 percent with mice, and, wait for it, 40 percent with bananas. And your genetic makeup is 99.9 percent identical to mine.

Darwin would be scratching his head, as this high level of genetic similarity seems a far cry from the immense diversity of life that inspired his famous theory. What's going on here? How similar are we, microscopically, to our closest surviving relatives in the tree of life? And how can we explain our manifest physical, behavioral, and sociological differences in the face of that sameness? Those questions puzzled two people in particular, a PhD student and her supervisor.

SAME, SAME BUT DIFFERENT

Eons ago, in 1975, Mary-Claire King and Allan Wilson were comparing the DNA of humans and chimpanzees,* using every technique available at the time. Proteins are what coding DNA codes for, via three-letter "words," or *codons*, that translate into one particular amino acid. A string of amino acids, like beads on a string, then form a protein. The four letters—the A, C, G, and T from the nucleotides that make up DNA—allow for a total of 64 possible combinations (4^3). For example, AAA codes for lysine, CAG for glutamine, GTT for valine. The system is a little sloppy, so multiple codons can stand for the same amino acid. Both TTC and TTT code for phenylalanine, for example. Thus, we say that TTC and TTT are synonymous because both sequences produce the same amino acid. (As an aside, redundancy in the system is yet another problem in using SNPs as genetic fortune-tellers; many differences have no biological effect.) The underlying assumption for the kind of comparisons King and Wilson relied on is that the greater the similarity in proteins between species, the more closely related those species are. We can use the same sort of comparison to estimate how long it has been since the species shared a common ancestor. The fewer accumulated differences between the species, the shorter the time since they separated from their last common ancestor.

---

* There are two species of chimpanzee and both are equally related to us: the common chimpanzee (*Pan troglodytes*) and the pygmy chimpanzee or bonobo (*Pan paniscus*). Throughout, I will use *chimpanzee*, although most studies probably will have used the common chimpanzee.

The first technique King and Wilson used directly compared the amino acid sequence of proteins that have the same function in both humans and chimpanzees, known as *homologous proteins*. Just like the codons can differ slightly but still produce the same amino acid, the same protein can have a slightly different amino acid sequence. And as with an actual DNA sequence, the assumption is that the more similar the amino acid sequence, the more closely related the species are.

King and Wilson's second technique was a more roundabout way of comparing certain proteins in chimpanzees and humans, using the immune response of rabbits after they were injected with those proteins. It is a neat trick to use the response of a rabbit's (or of any other vertebrate's) immune system when it's challenged with something alien. When our bodies are invaded by bacteria, viruses, fungi, or parasites, our immune system makes antibodies—proteins that latch onto the invader. Antibodies are invader-specific, so different kinds of, say, bacteria elicit the production of different antibodies. King and Wilson compared the rabbits' antibodies to serum albumin, a protein found in the blood of both chimpanzees and humans (and all other vertebrates). And once again, the more similar the antibodies produced, the more recent was the common ancestor of both chimpanzees and humans.

Their last protein comparison looked at the size of the proteins using gel electrophoresis, which is a way to force large molecules through a gel matrix using an electric current. The larger the molecule, the slower it moves through the matrix because larger things have a harder time finding their way through the gel matrix. And the more similar the molecules are in size, the more similar the human and chimpanzee proteins move in terms of speed. But there is more to this. Amino acids differ in their electrical charge and because gel electrophoresis uses an electric current to pull the molecules through the gel matrix, any difference in charge will produce a difference in travel speed. So, when two homologous proteins are not identical in their amino acid makeup, that shows up as a speed difference.

The techniques King and Wilson had available to them, plus the published data they included, are, by modern standards, considered crude, but they were the best for the time. Their conclusion, combining the results from their three separate techniques, was astounding: the average human protein is more than 99 percent identical to its chimpanzee counterpart. But they did not stop there.

The last comparison King and Wilson used is a neat way of looking at genetic similarities if you can't examine the complete DNA sequences directly (which we can do now, but not then). The DNA double helix is held together with bonds between the nucleotides, and those bonds can be broken with the application of heat, separating the two strands. With some ingenuity, it is possible to synthesize hybrid DNA molecules using one strand from a human subject and the other from a chimpanzee. The more similar two species are in their nucleotide composition, the more chemically stable this hybrid molecule will be. And the more stable the bonds, the more heat you need to break them. So, it is possible to measure similarity, via strength, by measuring temperature. That is, the difference in heat required to break the "pure" double helix and to break the hybrid molecule allows quantification of the similarity between the DNA from the two species.

This hybrid technique, unlike the methods that focus on

proteins, looks directly at the similarities in *all* the DNA in humans and chimpanzees. The distinction is important because we now know that genomes are full of what Dawkins memorably named "junk-DNA," or DNA with no known function. If junk-DNA really is junk—in other words, it does nothing for the organism—we expect such DNA not to be going through the process of natural selection. That is, any mutation that occurs in junk-DNA is "invisible" to natural selection because it doesn't do anything. And because they have no effect, such mutations can then accumulate, leading to more differences between the species than one would expect when looking at functional stretches of DNA that code for stuff like proteins. The longer two species have been separated, the more they will have garnered unique mutations. Because King and Wilson relied on studies that used slightly different techniques, they did not give a precise similarity percentage for this particular kind of comparison. What they did say was that the two genomes, measured in this way, are very similar.

By 2003, it was possible to compare DNA sequences more precisely than King and Wilson ever could, using methods developed during the Human Genome Project. And so a group of scientists compared the DNA sequences of 97 human genes to those gene counterparts in chimpanzees. These genes all coded for proteins. The result? Same as with the older techniques used by King and Wilson: overall 99 percent similarity.

Whether we like it or not, there is no denying that humans and chimpanzees are *at least* genetically as similar as sibling species in other groups of organisms. Different species of fruit flies from the genus *Drosophila*, for example, are as similar to each other as humans and chimpanzees are similar. We're more closely related to chimps than either of us are to gorillas, orangutans, or any other primates. By these standards, either chimps belong in our genus or we belong in theirs. If we aren't prepared to call chimps *Homo*, or human, it's not because of their DNA. Yet, we desperately hang on to our special place in nature, despite pervasive evidence to the contrary. To understand why we do this, we need to go back to the father of science itself: Aristotle.

The way we do science today has inevitably been influenced by the scientists who came before us. Darwin did his best to remove humans from their self-erected pedestal, but at the time he was a lone voice in the wilderness. Even after Darwin's great insight, many continued to approach our evolutionary history as if humans are something quite different from everything else. But in going back to where it all began, with Aristotle, I do not mean to ridicule those who came before us. Instead, I acknowledge their insights; after all, we are all products of the past, and none of us works in isolation.

OUR PLACE IN NATURE

Aristotle's *scala naturae* arranges animals on a scale of perfection. Humans are found at the top of the evolutionary scale, but interestingly, under angels and God. Even now we see this view represented in a common cartoon version of human evolution, one that irritates me immensely: a chimpanzee-like creature morphs into a fierce human hunter, first via a hairy half-chimp walking on two legs and then as a less hairy human-like version holding a stone tool. The assumption is that our common ancestor with chimpanzees looked like a chimpanzee, and that chimps stayed more or less the same throughout this time following our split.

We, humans, on the other hand, made a great leap forward. See, we are special!

Surprisingly, the great Thomas H. Huxley, the biggest defender of Darwin's theory of evolution at the time, was himself guilty of a gross misrepresentation of human evolution. In his 1863 book *Evidence as to Man's Place in Nature*, Huxley used a drawing by Benjamin Waterhouse Hawkins of the skeletons of a gibbon, orangutan, chimpanzee, gorilla, and man (in that order) to illustrate the progression from beast to man. The scale was adjusted so it was abundantly clear that each succeeding form morphed from the former one, with the "primitive" forms slightly leaning forward, ready to change into the next one. Perhaps in response to Huxley's misrepresentation, Darwin wrote in his second most influential book published after Huxley's, his *Descent of Man, and Selection in Relation to Sex*: "But we must not fall into the error of supposing that the early progenitor of the whole Simian [apes and monkeys] stock, including man, was identical with or even closely resembled any existing ape or monkey." I suppose that the creators of the cartoons don't read Darwin.

Aristotle organized organisms according to their broad similarities (being able to move or not, giving live birth versus laying eggs, having blood or being "bloodless"), but of course he did not know why some organisms were more similar to each other than to others. Neither did the most famous organizer of things, the eighteenth-century Swedish naturalist Carl (Carolus) Linnaeus. Linnaeus systematically grouped organisms based on their shared morphological characteristics. Initially he devised three Kingdoms: a Kingdom for things that move (the animals or Regnum Animale); a Kingdom for things that do not move (the plants, or Regnum Vegetabile); and a Kingdom for things not considered to be alive (the minerals, or Regnum Lapideum).

It was Linnaeus's goal to describe and name all species on earth. It goes without saying that he failed miserably. In 2009, a total of 19,232 new species were described—about twice as many as were known in Linnaeus's lifetime. Linnaeus certainly gave it a good try, though, and collected specimens whenever and wherever he traveled. But his own travels were insufficient to get him specimens from around the world, so he enlisted the help of loyal and keen students, who became known as his "apostles." Many of those apostles traveled the world as part of the Swedish East India Company. That sounds romantic, but travel in the eighteenth century was far from comfortable or safe. Seven of seventeen apostles died during their travels, a mortality rate of almost 40 percent. When one of his married apostles died of malaria, his wife, not unreasonably, blamed Linnaeus for making her children fatherless. Being a considerate man, Linnaeus from then onward sent only unmarried men on collecting expeditions.

Linnaeus was the first to group humans with other apes and monkeys. This group, the Anthropomorpha, included humans, monkeys, apes, and—wait for it—sloths. Not that Linnaeus had actually seen nonhuman apes and was struck by their similarity to us. He had only seen drawings of apes included in a dissertation by one of his pupils, Christianus Emmanuel Hoppius. According to Huxley, Hoppius's drawings are laughably inaccurate, showing "apes" that look like very hairy humans. In later editions of his book *Systema Naturae*, Linnaeus replaced Anthropomorpha with Primates. To this day, we still use the Primates classification, but we did take the sloths out of the group (sloths are most closely related to anteaters).

Linnaeus's aim was to create an ordered view of the natural world, and his order was based on morphological similarities. I relate to his desire for order; when I was a child, I liked to sort things by their size. What I sorted could be anything, really—stones I collected on my walks (which my parents needed to carry for me), shells from the beach, my stuffed toys. It seems that I passed on my childhood obsession to the next generation, as one of my daughters cannot eat lollies (also known as candy) without first sorting them by color. Of course, there was no reason to sort things by their size, nor is there any logical reason for my daughter to sort her lollies by color, but to our minds this sorting creates order.

In taxonomy, the grouping of organisms is based on the number of characteristics they share. While my daughter and I use only one sorting characteristic—color or size—biological classification systems use as many characteristics as are deemed informative. The more characteristics two organisms share, the lower the category they are placed in together. So, a *species* represents a group of organisms that share the most characteristics. But it wasn't until Darwin that we started to understand where this order in the natural world comes from: shared ancestry.

When done properly, a biological classification system tells us something about evolutionary relationships. That is, organisms share certain characteristics because they share an evolutionary history. The more recent their common ancestor, the more characteristics those organisms share. While we now use DNA sequences to compare species, the first biological classification systems could be based only on morphological similarities. But even with DNA, it is an art to determine the boundaries between species. Mother Nature does not easily comply with our desire to strictly delineate her creation.

Thanks to DNA, we now know that we last shared an ancestor with chimpanzees 6 to 7 million years ago. Alfred Russell Wallace, the co-discoverer of evolution by natural selection, was probably the first to refer to the human as "a naked ape." But as we'll see in later chapters, many of our ancestors who made an appearance after our split from the common ancestor with chimpanzees were also naked. "Talkative ape" is probably a more fitting description of our species. After all, humans are the only ape that regularly sits down to dwell on its own amazingness. And for that, it was thought, we first needed to evolve a large brain. As is so often the case, Darwin set the scene.

In *The Descent of Man,* Darwin wrote, "The difference in mind between man and the higher animals, great as it is, is certainly one of degree and not of kind." Language "owes its origin to the imitation and modification, aided by signs and gestures, of various natural sounds, the voices of other animals, and man's own instinctive cries." In Darwin's view, language was not a human-specific instinct, as some of his contemporaries claimed, but rather had to be learned, just like birds need to learn the specific songs of their species. It is the ability to learn that has evolved by natural selection. Darwin even dreamed up scenarios in which "ape-like animals" could have used sounds to convey information to their fellow beasts. With that increased use of sound, Darwin hypothesized, the vocal organs would have been "strengthened and perfected" until the lack of mental powers prevented any further evolution of speech.

Darwin's last phrase struck a chord with many people. To

be considered human, a creature must have a brain large enough to allow it to speak. And if Darwin's theory of evolution by natural selection was right, then there should be a gradual change in brain size, from small-brained chimpanzee-like creatures to human-like large-brained ones. The idea of the "missing link" was born, the search for which became at least one person's lifelong obsession.

IN SEARCH OF THE MISSING LINK

Despite his French-sounding name, Marie Eugène François Thomas Dubois was, like me, born in the Netherlands. He lived in the small village of Eijsden, the son of an immigrant apothecary from Belgium. The pharmacy was a great place to hear the village's news and gossip, and so a ten-year-old Dubois got wind of an upcoming lecture by the German biologist Karl Vogt about Darwin's theory that all living things share a common ancestor, including humans. Prevented from attending the actual lecture by his father, the young Dubois had to make do with the snippets he heard from the townsfolks and the newspaper reports. Vogt's lecture had caused enormous uproar among the burghers of Eijsden. He claimed that humans were not especially created by God but, rather, are a "mere cousin to the savage and bestial apes." Blasphemy! But for Dubois it was the beginning of a lifelong passion.

Dubois was encouraged by his receptive science teacher, who lent him Darwin's *Origin of Species* and *Descent of Man*, Ernst Haeckel's *History of Creation*, and Huxley's *Man's Place in Nature*. Haeckel's book in particular had a huge influence on the young Dubois, with his writings about the "Ape-like men," or Pithecanthropi, who must have evolved from the "Man-like Apes," or Anthropoides. The Pithecanthropi, Haeckel claimed, were the missing link between ape and human because they "lacked one of the chief human characteristics, articulate speech and the higher intelligence that goes with it, and so had a less developed brain." They, the Pithecanthropi, would represent a transitional form between apes and humans. The young Dubois was going to find this missing link—*Pithecanthropus*—and so prove, without a doubt, that humans evolved from other apes. Take that, burghers of Eijsden!

A good anthropologist knows how bodies are built, and so, after finishing his medical degree at the University of Amsterdam, Dubois accepted a position as anatomy instructor at the same university to study the evolution of the larynx—the vocal cord, or voice box. That's a research avenue I would have heartily supported, but I can't really imagine him as my student. By all accounts, Dubois was prickly, a characteristic that caused him many troubles when dealing with some of his big-ego contemporaries.

The German pathologist Rudolf Virchow was one such big ego; Virchow once boasted that he *was* German science. Regardless of his unpleasant character, Virchow was a scientific force to reckon with, known as the founder of modern pathology and the discoverer of the cause of cancer. To his colleagues he became the "Pope of medicine." But evolution, Virchow did not understand, calling Darwin an ignoramus and his theory of evolution an unproven hypothesis. So, when in 1887 the geologist Max Lohest and the anatomist Julien Fraipoint published a massive tome describing the fossil finds of two Neanderthal skeletons, or *Homo neanderthalensis*, Virchow was less than impressed. The discovery reignited an old debate about the place of Neanderthals in human evolution. Were they an example of an ancient, primitive race from which modern humans evolved? Perhaps. Or, alternatively, were they our contemporaries, our cousins? What was undisputed was that the fossils were too "human" to be the link between humans and nonhuman apes. Because they were so human, Virchow saw them as evidence for the *absence* of evolution, claiming—after having studied a Neanderthal fossil himself in 1872—that it did not represent a primitive form of human but, rather, was an abnormal human being that, judging from the skull, had been malformed or injured. The fossil's unusual bones must have been the result of diseases like arthritis and rickets. When Virchow spoke the following words, "The intermediate form is unimaginable save in a dream," Dubois took them as a call to arms. And so an unknown anatomy instructor set off to prove wrong the giant of German science.

Dubois abandoned his longstanding project on the evolution of the larynx and his fights with Max Carl Anton Fürbringer (his superior and another big ego), and in 1887 he packed up his belongings and with his pregnant wife Anna and young daughter Eugenie, joined the Dutch East India Company as medical officer. Haeckel had argued that humans originated in Asia, and for a Dutch burgher, the East Indies was a logical destination. After many failed attempts and bitter disappointments, and multiple relocations of his growing family, Dubois's dogged persistence finally paid off. In 1891, in a bend of the Bengawan Solo River, near the tiny village of Trinil, on the island of Java, Dubois found what he had been dreaming about his whole life. It was a cranium, a tooth, and a femur that had belonged to a creature not yet quite human but no longer ape. Dubois had found his missing link, *Pithecanthropus erectus*, or Java Man.

But had he?

PITHECANTHROPUS ERECTUS BECOMES HOMO ERECTUS

In 1927, Davidson Black, a Canadian medical doctor working at the Peking Union Medical College in what is now Beijing, figured that fossil remains found in China belonged to a completely new group. He gave the fossil the name *Sinanthropus pekinensis*, or Peking Man. Three more *Sinanthropus* species would later be identified, all named after the places where they were found in China: *S. lantianensis* (Lantian Man), *S. nankinensis* (Nanjin Man), and *S. yuanmouensis* (Yuanmou Man). Black argued that his *S. pekinensis* was the missing link, not Dubois's *Pithecanthropus erectus*. He was definitively on to something, but not quite what he thought he was, as we'll soon see. Then, in 1936, the Dutch-German paleontologist and geologist Gustav Heinrich Ralph von Koenigswald found yet another fossil on Java, near Mojokerto. Von Koenigswald initially gave the fossil the name of *Pithecanthropus modjokertensis*, but he soon got into immense trouble with Dubois, who was fiercely protective of his own *Pithecanthropus*. To avoid the wrath of Dubois, whose prickly nature had not mellowed, von Koenigswald renamed *P. modjokertensis* as *Homo modjokertensis*.

Homo, first introduced by Linnaeus, means "human being" or "man," in the general sense. The fact that von Koenigswald could easily decide that "his" fossil should be renamed so as to belong to a completely different group shows how ill-defined the human and human-like species were then. From the late nineteenth to the mid-twentieth centuries, even more genus names popped up in addition to *Pithecanthropus* and *Sinanthropus*. *Proanthropus, Cyphantorpus, Africanthropus, Telanthropus, Atlanthropus,* and *Tchadanthropus* all referred to human-like fossils found throughout Asia. The seemingly immense diversity of human-like species in Asia appeared to confirm the widely held belief that modern humans evolved in Asia. But how real was this diversity, given how arbitrarily paleoanthropologists—people who study human evolution mainly from fossils—dole out new names? Enter one of the twentieth century's leading evolutionary biologists.

German-born Ernst Walter Mayr was a jack-of-all-trades. One often thinks of a jack-of-all-trades as someone who dabbles in many skills with no expertise in anyone field. Not Mayr. An avid birder, Mayr excelled in systematics and taxonomy, and not only of birds; he was also a philosopher of biology and a historian of science. Then in 1950, Mayr decided to get involved in the systematics of humans and their ancestors. Bewildered by the diversity of names, he decided to subsume them all within three species of *Homo*: *H. transvaalensis, H. erectus,* and *H. sapiens*. Although what he called *H. transvaalensis* later became the genus *Australopithecus*, all species of human-like fossils found in Asia to this day remain lumped together under the name *H. erectus*. That means that Dubois had not found the missing link. Nor had Black. Both had found the fossil remains of *H. erectus*. Dubois died before it became clear that his missing link was not what he thought it was. It would have broken his heart to know he failed his life's mission.

Homo erectus was the first species belonging to the same genus as us modern humans to have left Africa, the true cradle of humanity. Once outside Africa, *H. erectus* diversified into multiple distinct species of *Homo* across Europe and Asia, and continued to roam the earth well after our own species appeared. Then, it seems that some *H. erectus* individuals got stuck on an island. On the island of Flores, to be precise.

DWARF SISTER OR DISTANT COUSIN?

In 2003, in a cave on the Indonesian island of Flores, fossil remains were found of what appeared to be a tiny human-like creature. The team of scientists who described the remains concluded that the finds belonged to a previously unknown species of human. They called it *Homo floresiensis*. Because of its diminutive size, only about a meter (3½ feet) tall, *H. floresiensis* quickly became known as The Hobbit, a nod to the immensely popular *Lord of the Rings* films at the time.

The Hobbit skeleton was relatively intact, with a fairly complete head, a leg and part of the other leg, and bits and pieces of the rest. Even in 2003, it was still difficult to classify human remains. In fact, nothing much had changed since Dubois and his contemporaries. To figure out where a fossil fits within our evolutionary history, paleoanthropologists work backwards. Using features that distinguish us from chimpanzees, they look at fossil remains and determine how far they are on their way to being like us. Any human remains that do not fit within present-day human variations are referred to as "archaic."

Paleoanthropologists are obsessed with three characteristics. Does the creature walk on two legs? How big is its brain?

And how small are its teeth? Teeth can tell us something about the creature's diet, but they also can hint at how the creature dealt with rivals, whether using its teeth ("primitive") or weapons ("derived") to fight. Primitive characteristics are, then, those that are furthest removed from modern humans—us—while the derived are those that most resemble those characteristics found in us. Dubois's fossil had a brain size greater than modern chimpanzees but smaller than modern humans, while its femur indicated it walked on two legs. The tooth, a molar, could easily have been from a modern human or from a nonhuman ape; it had characteristics of both.

Homo floresiensis in some ways resembled the more archaic human known as *Australopithecus afarensis*, better known as Lucy, that was found in Ethiopia in 1974 (named *H. transvaalensis* by Mayr; the name Lucy came from the 1967 Beatles song "Lucy in the Sky with Diamonds"), while in other ways the skeleton showed similarities with other humans classified as *Homo*, such as *H. erectus*. Size-wise, *H. floresiensis* was more similar to Lucy but lacked its tooth and jaw structure. Its face was more similar to *Homo* than to *Australopithecus* because its skeleton showed no evidence of a protruded face, as in *Australopithecus* and species that came before. From its skeleton, it also became clear that The Hobbit walked exclusively on two legs, another hallmark of humans and one much easier to determine than the ability to speak. It also looked as if The Hobbit had made tools; scattered around the skeleton were stone artifacts and animal remains. *Australopithecus afarensis* or other species of archaic humans outside the genus *Homo* are not known to have made tools. Ergo, The Hobbit had to be part of our group, *Homo*.

There was something peculiar about *H. floresiensis*, though. It had a remarkably small brain, even when corrected for body size. Its small brain made other paleoanthropologists doubt the skeleton represented a new species. Instead, they argued, The Hobbit was simply a deformed modern human, *H. sapiens*. Others, who reanalyzed the fossil remains and compared them to modern humans suffering from a disease that results in a tiny brain (microcephalia), concluded that The Hobbit really is a new species. *Homo floresiensis*, according to them, is the end product of a long period of isolation of *H. erectus* on the small island of Flores. Insular dwarfism is a well-established evolutionary process whereby originally large-bodied organisms confined to a small island become much smaller because of their much smaller environment. The Hobbit thus seems very likely to be one of our distant cousins, albeit a very small one.

Homo erectus gave rise not only to *H. floresiensis* but also to *H. heidelbergensis,* the first species of *Homo* thought to be well adapted to colder climates. Remains belonging to *H. heidelbergensis*—an almost complete jawbone—were found near the southern German city of Heidelberg in 1907. Illustrating the difficulties of deciding what is what, especially when one has only a jawbone to go by, *H. heidelbergensis* was initially lumped with *H. erectus*, but is now widely considered to be its own species. Then, around 2010, when researchers were able to sequence the genome of some fossil remains, it turned out that *H. heidelbergensis* gave rise to two new species, one in Africa and one in Europe. The latter became *H. neanderthalensis* and the former *H. sapiens.*

The ability to sequence DNA from human fossils opened

up a whole new way of looking at our past. And that past became even more interesting. Our DNA is able to tell a story our fossils never could.

A REVOLUTION IN OUR EVOLUTION

When Svante Pääbo arrived at the Max Planck Institute for Evolutionary Anthropology in Leipzig, Germany, in 1997, to solve the mystery of human evolution, he had an extensive molecular toolkit at his disposal. Originally from Sweden, Pääbo had been interested in our ancestry ever since he was a PhD student when, instead of studying viruses as he was supposed to do, he decided to try to extract DNA from a two-thousand-year-old Egyptian mummy. Being ridiculed for wasting his time on a shriveled corpse did not deter Pääbo, who was convinced that if he could extract DNA from old human remains, he would be able to reconstruct our evolutionary past. One man shared this young PhD student's vision—noted biochemist Allan Wilson, who was impressed with the mummy work when he read the publication. Not realizing that Pääbo did not even have a PhD, Wilson asked if he could come for a sabbatical so they could together develop ancient DNA work. Instead, Pääbo moved to Wilson's lab at the University of California, Berkeley, in 1987, where the two worked on the genomes of extinct mammals. Sadly, Wilson died young, in 1991.

When Pääbo entered the field of human paleogenetics, it was possible to successfully extract DNA only from remains that were at most a few thousand years old. That all changed in 2021, when a group of scientists successfully extracted DNA from a tooth of a mammoth, found in Siberia, that was more than one million years old. But with that innovation still in the future, Pääbo had to make do. And make do he did. He went on to transform the way we study human evolution. In 1963, Emile Zuckerkandl and Linus Pauling had coined the term *paleogenetics* to describe a hypothetical means for reconstructing a species' evolutionary history by looking at the sequence of its DNA. Pääbo's work turned that possibility into reality. Deservedly, in 2022 he received the Nobel Prize in Physiology and Medicine.

Pääbo's immediate goal when he arrived in Leipzig was to sequence the complete genome of the Neanderthal mitochondria (the powerhouses of the cell); his aim was to look for evidence of interbreeding between *H. neanderthalensis* and *H. sapiens*, which could be inferred if they found the mitochondrial sequence of a modern human in a Neanderthal fossil, or vice versa. I haven't asked him personally, but I strongly suspect Pääbo and his team were disappointed when they found no evidence of such interspecies sexual liaisons.

Mitochondria are only ever transmitted by the mother. The father's mitochondria are mostly found in the tail of the sperm cell, generating the energy needed for the sperm cell to swim. When the sperm cell enters the egg, its tail is left behind, and thus also most of the father's mitochondria. If some mitochondria do happen to get into the egg, they are quickly recognized by the machinery of the egg cell and destroyed. That means we all carry our mother's mitochondria, which came from our grandmothers, and before that from our great-grandmothers, and so on, ad infinitum. Using the mitochondrial genome, all living humans can be traced back to their *matrilineal most recent common ancestor* (MRCA)—our Mitochondrial Eve, who lived about 160,000 years ago in Africa.

"Mitochondrial Eve" does not refer to the first woman of our species, nor does it imply she was the only woman living at that time. Mitochondrial Eve, whom Allan Wilson preferred to call "Lucky Mother," simply had the luck that her female offspring went on to produce more female offspring, probably at a time when the number of humans was relatively small. In a similar vein, we can identify Y-chromosomal Adam because Y chromosomes are only transmitted from father to son.

Pääbo and his team established that modern humans do not carry Neanderthal mitochondrial genes, because we have no recent female Neanderthal ancestor. But nuclear genes would soon tell a different story. The genes in our nucleus are a mix of genes inherited from mother and father. And in those genes a researcher, sifting through the sequence data available online, found something odd.

David Reich is a geneticist at Harvard Medical School, and he was interested in using the data generated by Pääbo's group to compare the nuclear genome sequences from Neanderthals with those from modern humans. Much to his surprise, Reich found that some Neanderthal genomes were more similar to some humans than to other humans. To be precise, people from Asia and Europe shared some of their sequence with Neanderthals, whereas people from Africa never did. The evidence Pääbo did not find when he was looking into mitochondrial DNA Reich did find in nuclear DNA. Modern humans and Neanderthals interbred, but only outside Africa, as Neanderthals never lived in Africa. If you are of Asian or European decent, chances are 1 to 4 percent of your genome is Neanderthal DNA.

Then someone found a more than 40,000-year-old pinky in a cave in Siberia.

Initially, Pääbo and his team assumed that the pinky belonged to a modern human or a Neanderthal, and they didn't think much of it. But small as it was, this pinky held a surprise. Its DNA did not match the DNA from modern humans, nor that from Neanderthals. The pinky belonged to a different species altogether: the Denisovans, named after the Denisova cave in which the pinky was found. (Researchers are reluctant to give the Denisovans a proper scientific name, such as *Homo denisova*, because we know them only from their DNA.) The pinky yielded a lot of DNA, allowing a comparison between Denisovan and modern human DNA. Knowing what we do now about Neanderthals and modern humans, it probably doesn't come as a surprise to learn that the Denisovans, too, interbred with modern humans. People native to many Pacific islands carry 4 to 6 percent of Denisovan DNA sequences. Like their Neanderthal cousins, the Denisovans never lived in Africa, but they evolved, and went extinct, in Asia.

◆

Charles Darwin had no fossils and no DNA to guide his ideas about our evolutionary past—with one exception. While spending time in London, recovering yet again from a bout of illness, Darwin was shown an unusual fossilized skull that had been found in the Forbes Quarry in Gibraltar in 1864. The skull was being prepared for display at a meeting of the British Association for the Advancement of Science, in Bath, by George Bulk, an English paleontologist, and Hugh Falconer, a Scottish paleontologist. Prior to the meeting, Falconer had brought the skull to his friend, Charles Darwin. It seems that, unusual for the great man, Darwin was not impressed. No other writing by him mentioning the skull exists other than a throwaway line in a letter to his friend Joseph Dalton Hooker, mainly describing Falconer's visit. The skull was that of a female Neanderthal. Even intellectual giants miss something sometimes. And what if Darwin had known about the 400,000-year-old fossil remains of *Homo heidelbergensis*, discovered in 1935, but found nineteen miles from his house? We'll never know.

After this briefest history of the study of our evolutionary past, it is now time to return to 1975. That's because Mary-Claire King and Allan Wilson had something specific to say in their paper.

MORE THAN GENES

Ignored by all but the most careful readers, King and Wilson didn't mean to focus on the close genetic similarity between us and chimpanzees. Instead, they were intrigued by the concept that very similar genomes can produce very *different* organisms. They mused about possible mechanisms that could explain our morphological and behavioral differences. Many of their guesses turned out to be correct. Yet, it remains tempting to try to solve the mystery of the 1 percent difference between us and chimpanzees—the yawning gulf between genotype and phenotype—by scrutinizing the assembly instructions encoded in that tiny fragment of DNA. For any macroscopic trait of interest, we can always ask: Is there a "snip" for that? The "selfish gene" has been a tremendously successful meme. Remember the headlines? Scientists find the gay gene, the gene for schizophrenia, even the language gene (allegedly *FOXP2*). None of these claims have held up, but there's still a cottage industry in evolutionary biology that is devoted to finding a single gene for X, Y, or Z.

The genomic revolution in biology—alongside parallel work in anatomy, anthropology, and archeology—is revealing that a precious few key changes made for the leap forward from our last common ancestor with chimpanzees. We are the result of a short series of mistakes that made us what we are today. A few of the flukes that now define us were indeed genetic. But the gene's-eye view of natural selection has sometimes led us to overlook simple, physiological explanations for our differences from other apes. The American philosopher Hilary Putnam once pointed out that you might try to explain why a square peg won't fit in a round hole by using quantum mechanics—but why bother, when a geometrical explanation works just fine?

At the end of the day, the key to our evolutionary success is to raise offspring who can thrive, raise children of their own, and so on down the line. The simple calculus of natural selection, operating simultaneously at macro- and microscopic scales, has given rise to the immense diversity of life we see today. That diversity ranges from single-celled organisms to giants like the sequoia tree and the blue whale, with everything imaginable in between. It even includes creatures that challenge the imagination, like the male anglerfish, which has evolved into an appendage of the female. All these species—the products of millions of years of accumulated changes—managed to leave behind more offspring than other species. And so did we.

two

OUR ORIGINAL CHILDCARE PROBLEM

In the summer of 2014, on a quiet cul-de-sac in suburban Oak Ridge, New Jersey, a black bear was caught on video sauntering around the neighborhood *on two legs*. At first, people thought it was a hoax. But it turned out that Pedals, as he came to be known, was indeed bipedal as a result of catastrophic damage to his front paws. The footage shows him trying, once in a while, to get back on four legs, but almost immediately jolting upright, presumably in pain, and reverting to an eerily anthropomorphic gait. The resemblance was so uncanny that some commentators insisted Pedals *had* to be a person in a bear suit.

But Pedals was real—a vivid illustration of the biological fact that, when the need arises, skeletal structure and musculature can adapt quickly to accommodate changes in locomotion. Permanent paw damage meant a permanent change in posture, requiring significant adjustments to the pelvis and leg muscles. Of course, these developments wouldn't be passed down to offspring—and not just because Pedals was killed by hunters two years later. Evolution isn't that Lamarckian.

Pedals became an internet sensation because bears don't normally walk on two legs, but every parent has watched their child make the transition from crawling to walking, getting better with time and practice. Our children, and poor creatures like Pedals, make the change from life on all fours to bipedalism thanks to phenotypic plasticity—the ability to adapt to changing conditions during the lifetime of an individual. Plants are masters of phenotypic plasticity for the simple reason they cannot pack up and leave to move somewhere else; instead, their best bet is to make do. Phenotypic plasticity is one reason why plant taxonomy is a complete nightmare, reserved for only the most determined (or insane?). Two plants within the same species can look nothing alike, even when they are clonal and thus have identical genomes. But my favorite plant example surely is the vine that has as its scientific name *Boquils trifoliolata* (not all organisms have a common name, sadly). This vine is capable of mimicking the shape of the leaves of the plants it lives next to, or sometimes on. Why? No one knows really, although there is a lot of speculation. But let's get back to our children, and by extension, to Pedals.

The skeletons of both Pedals and our children undergo major changes once they start walking instead of crawling—or in the case of Pedals, walking on two legs—because of the change in muscle movement that comes with a change in posture. Strange as it seems, perhaps, muscle contractions affect the *regulation of genes* that are key to the construction of bone and cartilage. (One reason weight-bearing exercise is good for bone strength.) It is then easy to imagine how a simple change in the way one moves affects the shape of the structure required to allow such change in movement. It's a nice example of positive feedback: the more one walks on two legs, the better one becomes adapted to doing so.

In a way, the transition both Pedals and we made when we were very young mimics what happened to our ancestors over evolutionary time. As they gradually moved out of the trees, first spending some time on the ground, then more time to eventually permanently, their skeletons adapted. That gradual change we can trace back in the fossil record. And that fossil record, bone by little bone, tells us that our predecessors started to leave the trees a very long time ago—much longer than anyone had anticipated.

At least 4.4 million years ago, apparently. At that time, one of the earliest species of hominin—of the family Hominidae, which includes modern humans, extinct human species, and all our immediate ancestors—roamed what is now the hottest place on earth. It is so hot that the Awash River's water flow evaporates as quickly as it arrives. Here in Ethiopia's desolate Afar Rift, an Ethiopian student with a gift for finding fossils, Yohannes Haile-Selassie, in 1994 found a piece of a hand bone that turned out to belong to a fossil that became known as Ardi, a beautifully preserved and fairly complete skeleton of a female *Aridipithecus ramidus*. It took almost fifteen years for Ardi to be safely extracted and preserved. Ardi's discovery changed everything we thought we knew about our past.

Ardi was clearly a very early hominin but did not look like a chimpanzee. Instead, she had features of both chimpanzees and humans; she could walk on two legs, had feet adapted to walking, but still had the flexible big toe needed to grasp tree branches. She climbed trees in which she moved about on all fours, but she lacked any sign of knuckle walking typical of chimpanzees and gorillas. Her hipbones were human-like in some respects, but ape-like in others. Ardi also had small canine teeth. Her discovery shattered the long-held belief that humans descended from a chimpanzee-like ape (remember that cartoon I dislike so much). Chimpanzees and gorillas do not represent our primitive ancestors; the characteristics that both share, like knuckle walking, are not leftovers from the common ancestor we all three once shared. Instead, chimpanzees and gorillas adapted to similar conditions, just like we adapted to ours. Ardi, with her mixture of ape-like and human-like characteristics seemed to fit the description of the missing link. Trouble is, there was something quite wrong with Ardi: Her brain size did not make the cut. Common wisdom had it that the road to modern humans started with a big brain. And for that there was good evidence. Or was there?

In 1912, in the village of Piltdown, Sussex, UK, the amateur archeologist Charles Dawson found an exceptional fossil that became known as Piltdown Man. Piltdown Man had a brain about two-thirds that of a modern human but a jaw that was chimpanzee-like with a tooth that could have been from either a human or a chimpanzee. The missing link found on British soil! An ancient Englishman, with a massive brain and the face of an ape! Just what we had been looking for. Piltdown Man was estimated to be about 500,000 years old, and so the claim of his being an intermediate form between apes and humans seemed reasonable. In part owing to national pride (to date, important fossils had been found only in Germany, Belgium, and France), the authenticity of Piltdown Man was only too readily accepted by British scientists and the public alike. Piltdown Man was even more exceptional because Dawson had also found primitive tools with the fossils, suggesting that Piltdown Man had the mental capacity and dexterity to craft them. The fossil, which received the scientific name *Eoanthropus dawsoni*, seemed a good alternative to the small-brained fossil Marie Eugène Dubois thought was the missing link.

It wasn't until 1953 that the world found out that the fossils were fakes. The "first Englishman" was a compilation of a human skull and an orangutan jaw and tooth. Its "tool," which reminded people of a cricket bat (someone had a sense of humor), was likely a fossilized elephant bone recently carved with a steel knife. Regardless of one's sense of humor, the consequences of the hoax were significant. It allowed many to ignore real fossils that told us something important about our past—fossils such as the Taung Child.

The fossilized skull that would become known as the Taung Child was found in South Africa in 1924. It was sent to the Australian anatomist and anthropologist Raymond Arthur Dart, who had just moved to Johannesburg to take over the anatomy department at the newly established University of Witwatersrand. Dart was interested in some baboon fossils that had surfaced in a lime mine at Taung, in the Northern Cape of South Africa. Two boxes with fossils were delivered to his home one Saturday afternoon when Dart was supposed to be getting dressed to serve as the best man at a wedding. *Just a quick peep into the boxes*, he thought, before we would head off. The first box was a disappointment, but the second one almost made him forget the wedding. Dart picked up the cast of a brain, three times as large as that of a baboon and bigger than that of an adult chimpanzee, but smaller than that of a modern human. No one had ever seen a brain like the one Dart was holding in his hand.

By Christmas 1924, Dart had prepared the fossilized skull and given it a name: *Australopithecus africanus*, the "southern ape from Africa." The Taung Child had a brain and face the shape and size of which were not yet human, but at the same time were no longer ape. And because it clearly walked on two legs, it showed an intriguing mix of ape- and human-like characteristics. The Taung Child was much more primitive than both Dubois's *Homo (Pithecanthropus) erectus* and the Neanderthals known at the time. (Ardi had not yet been found.)

Dart's contemporaries dismissed the Taung Child as a relative of a chimpanzee or gorilla, based on its small brain, and being nothing of real interest. And so, thanks to a fake fossil and people's willingness to believe in a constructed reality, *Australopithecus africanus* was more or less ignored until a famous sister entered the scene in 1974. We've met the sister before. Remember Lucy, *Australopithecus afarensis*? One fossil from Africa can be dismissed as an aberration, but surely not two. Finally, the search was on for our African ancestry; a fruitful search it was, too, as Darwin had been right all along. Humans evolved in Africa, not in Asia. And certainly not in the UK.

To this date, more than twenty early human ancestor species have been identified from fossils found in various parts of Africa, some as old as 6 to 7 million years. Most of them are known only from jaw fragments and teeth, which indeed questions how one can determine if the specimen is from the human or chimpanzee lineage. If it has small, blunt teeth and evidence of bipedalism, we welcome it into our family. If you have the correct part of the skull, you can tell if the owner of the skull was capable of walking on two legs. Based on teeth, proficiency of bipedalism, and brain size, paleoanthropologists currently recognize three broad groups of ancestors, depending on their ages.

The earliest are older than 4.2 million years and include species such as *Sahelanthropus tchadensis* (6–7 million years) and *Orrorin tugenensis* (5.6–6.0 million years). Both species were bipedal, had teeth unlike those of apes, and an estimated brain size similar to that of modern chimpanzees. And then, of course, there's Ardi, *Ardipithecus ramidus*, but also a relative of Ardi, *A. kadabba*. Lucy and her sister species, all species of *Australopithecus*, lived between 4.2 and 1.9 million years ago. Theirs was a successful group, diversifying into many species. How many we don't know because of the difficulty in deciding what a new species is, especially if the only thing we can rely on are some bits and pieces of skeletons. What experts do agree on is that one of the *Australopithecus* species gave rise to the first member of the next—and last—group, *Homo*, about 2.8 million years ago. That last group can be recognized by a larger brain, the absence of any features that would allow it to easily get up into trees, much smaller teeth than *Australopithecus*, and finally, tool making.

Since the split from the common ancestor with chimpanzees, our ancestors experimented with living partially in the trees and partially on the ground, until they committed to full-time bipedalism. It must have been a great time for experimentation, but for most species the experimentation turned out to be a dead end. For those who hung in there—or you might say, stood their ground—more changes were about to happen. Changes that ultimately would lead to our original childcare problem. And that all started with hips.

THE MALLEABLE PELVIS

If you look at your cat or dog, or any other animal that walks on all fours, you will see that their body weight is always supported by at least two legs. But when you have only two legs to walk on, with every stride one leg will need to support the body's full weight. Unless you are a toddler in the process of trying to walk (in which case you wouldn't be reading this), you do not realize how tricky it is to maintain your balance. Every time you put your full weight on one leg, the pelvis has the tendency to tip toward the unsupported side. One way to compensate for this imbalance is to lean your trunk toward the unsupported side, or to stretch your arms. That is exactly what a chimpanzee does when it (briefly) walks on two legs. But such side-to-side weight shifts are energetically costly.

We, modern humans, maintain our balance thanks to the lesser gluteal muscles, especially the gluteus medius, that cross laterally over the hip so they become abductors (a group of muscles that keep the thighs together) rather than extensors (muscles that extend limbs). To create a surface for the abductor muscles to attach to, our iliac blades, or hip bones, changed their orientation so they flare outward, forming the characteristic bowl shape of the human pelvis. They also became shorter to lower the center of our mass, another adaptation to increase stability. And this also prevents the lower back, which became curved to assist in balancing the trunk over the pelvis, from getting trapped in between. I won't bore you with all the other changes to the pelvis that accommodate the change in muscle attachment; suffice it to say there are many. (If you ever had a leg or knee injury and went to a physiotherapist or sports doctor, they may have explained the complicated ways in which leg muscles wrap around the pelvis to ensure stability during standing and walking.) All these modifications allow us to balance our upper bodies while taking long, efficient strides.

Because the pelvis is so important for locomotion, it is easy to see how paleoanthropologists figured out which of our ancestors could walk on two legs, even if their pelvis did not yet look quite like ours. Of course, there are also clues, like the head, as I hinted earlier, but also more obvious ones, such as the shape and form of the hands and feet. Ardi spent time both in the trees and on the ground, given that she still had a grasping big toe. But Lucy's foot, ankle, knee, wrist, and hand no longer showed the hallmarks of a creature that lived in trees. That means that by the time Lucy and her kin came around, our ancestors had become committed two-footers. So, Ardi wasn't some weird aberration—hominins had been walking on two legs for millions of years, and way before their brains became larger.

One change that Mother Nature did not think through, however, when she adapted the pelvis to bipedal locomotion is the required change in the orientation of the birth canal. The shape of the lower half of the pelvis, together with soft tissues such as the pelvic floor muscles, dictates the shape of the birth canal. In chimpanzees, the birth canal is large, almost oval-shaped with the broadest part facing forward, away from the spine. That means that when a chimpanzee baby is born, it comes out of the birth canal without making an internal turn. (The same is true for all monkeys and other apes, other than humans.) The baby is born facing mom so that she can safely grab the back of its head and gently help it out.

Because of the change to the pelvis as our ancestors got more proficient at bipedalism, the "midwife's view" of the birth canal became more narrow. From Lucy's fossil remains we know that her birth canal had already become so narrow her babies probably had to make a turn to fit their shoulders through it. (I add "probably" because the issue is still debated in the literature and most likely will never be resolved.) One consequence of that turn is that babies are no longer born facing mom but, rather, facing away from her. And that means mom can no longer assist the baby's birth without the risk of breaking its neck.

Combine a difficult birth with the inability of mom to be of any help, and you can see the need for others to assist. And this implies some form of social life. Speculative, I know, as it is difficult to make sense of social relationships from fossil remains, but there are other hints that suggest Lucy was not doing it alone. If you have ever tried to keep up with your dog when it was in a good mood for running, you know there is no way you can match its speed, no matter how fit you are. If your dog is a greyhound, it can run at 43 mph. Compare that to the fastest human, currently Usain Bolt, who reached a maximum speed (briefly) of 27.78 mph. By becoming bipedal, we gave up the ability to run fast.

There is more. I'll never forget a particular human skull I saw in a natural history museum some years ago. It had two neat holes in it, most likely made by a saber-toothed tiger. Not only did Lucy lose her ability to outrun any large predator, but she also lost the competence to quickly climb into a tree to escape it (which, incidentally would not have worked for a saber-toothed tiger). Luckily for Lucy, saber-toothed tigers did not live in Africa, but you get the point. Other big predators did, though, and still do, and the likes of Lucy had limited means to defend themselves. Especially those carrying babies.

Which raises another baby issue. Lucy probably had lost most of her body hair, and without lots of body hair her baby would not have been able to easily grab hold of mom, so moms had to use their arms to carry them around. But even if Lucy and her kin were fully hirsute, carrying babies the "ape-way" (initially on mom's belly and later on her back) is incompatible with walking upright. And when carrying a baby, there is not that much else you can do with your hands. Sure, holding a little one feels good—I can personally attest—but it is also an unescapable chore. It's a chore that made its entrance about 4 million years ago. At that time, our ancestor probably was something like this: a smallish, more or less naked creature, no longer capable of quickly making a retreat into trees, unable to run fast, and burdened with little babies. In other words, a species that *had* to live near others to stand any chance of survival.

It is probably fair to say that human sociality had its origin in *Australopithecus*'s pathetic-ness.

Pathetic or not, the group of human ancestors colloquially known as the Southern Ape hung in there for 3 million years or so. Scientists continue to debate whether *Australopithecus* walked just like we walk today, or whether a more human-like gait appeared only with the first species of *Homo*. Whichever way, what we do know is that once *Homo* made an appearance, the pelvis started to change again, probably because our early family members became even better at walking on two legs. In a way, they had to.

The landscape started to change, from abundant forests to more open, grassy savannahs. It was a landscape easy to move through, but also one that brought greater exposure to the sun. Full disclosure again, because not everyone agrees, but it seems that the change in landscape caused by a change in climate selected for more efficient locomotion. And a way to get more efficient locomotion is to narrow the width of the pelvis relative to body height, to increase the length of the legs, and to increase the crural index ratio—the ratio of thigh length to leg length. A higher ratio increases the force that can be applied against the ground, making the strides more powerful. A narrow pelvis had another advantage; it led to a narrower upper body, and was more effective at eliminating excessive heat. That's important if you live in a sun-exposed environment.

But there was another reason why the pelvis changed: Brain size made a jump. Lucy's head could fit a brain of about 375 cubic centimeters. *Homo habilis* had a brain of about 500 cubic centimeters. That is still nothing compared with our brain, which measures, on average, 1,350 cubic centimeters. Still, it is quite a jump in brain size between *Australopithecus* and *Homo*. A larger head means babies with a larger head, and so the pelvis had to change again to allow the babies to easily fit through the birth canal. The pelvis became a little wider from the back to the front.

That one structure, the pelvis, had to accommodate three demands. An overall greater pelvic width was good for non-complicated births, while a narrower pelvis offered both better thermoregulation and superior locomotion. It seems our ancestor reached a compromise that worked well enough. And for a very long time, brain size and pelvis stayed more or less the same.

WHEN AND WHY OUR BRAIN BECAME BIG

Jamie is faced with a problem. The little treat she wants to eat is stuck in a box. To get it out, she needs to find a stone that fits through the hole in the top of the box so the weight of the stone will flip the platform on which the treat sits, thereby releasing it. The trouble is that the stone she needs is itself stuck in a tube for which she needs a stick to retrieve it. Sounds daunting? Perhaps, but not for Jamie, as she soon figures out the sequence of events needed to achieve her goal. Other times, Jamie will craft the tool she needs if nothing suitable is lying around, modifying it so it is fit for obtaining the kind of treat she is after. For example, when the treat sits in a bucket in a well, Jamie will find a stick she can turn into a hook and then use the hook to lift the bucket out of the well. Or, faced with a snack stuck in a small hole of a tree, Jamie will prune off and trim a leafy oak branch to just the right width to poke it into the hole.

By now you might have guessed that Jamie is not a human; she is a New Caledonian crow, famous for an intelligence that easily matches that of chimpanzees. New Caledonian crows are not the only nonhuman animal tool users and tool makers, but probably one of the most famous because we didn't expect birds to be that clever. But that's not all. Who would have thought that little critters such as ants use tools? Granted, they are not known to make tools, but they certainly use them if that makes getting food easier, or to annoy their enemies. Some species use stuff that is lying around, like bits of soil or leaves, to soak up liquid food and then carry the food-soaked tools back to the nest. Others use soil to block the nest entrance of neighbors, giving a different meaning to the phrase "neighbors from hell."

Before we knew anything about tool use in New Caledonian crows and in ants, David Greenbeard became famous as the first nonhuman tool user. David Greenbeard was a chimpanzee, the first of many studied by the world's foremost expert on chimpanzee behavior, Jane Goodall. Goodall, an English primatologist and anthropologist, started her studies in 1960, traveling to Gombe Stream National Park in Tanzania at a time when not many women, or men for that matter, would consider such a journey. She was the first to document the many behavioral similarities between chimpanzees and us, a tradition later followed by my fellow Dutchman Frans de Waal.

When David Greenbeard felt like a treat, he would find a stick, strip off its leaves, and use it to fish termites out of their nest, which he then licked off the stick, just like we would do to get the last of our milkshake off the drinking straw. David Greenbeard was not unique in his tool-making abilities. Jane Goodall found evidence everywhere, once she started looking for it. Chimpanzees use leaves as sponges to soak up water to drink, and use rocks as weapons and as hard surfaces to crack open foods like gourds so as to eat the soft fruit inside. It seems there are even cultural differences among different groups of chimpanzees in the kind of tools they make and for what purpose. Yet, chimpanzees never evolved a large-brained relative, which is what Charles Darwin expected. And as Jamie and the ants show, you don't even need hands to create tools.

Yet Darwin wrote in *The Descent of Man*,

> The free use of the arms and hands, partly the cause and partly the result of man's erect position, appears to have led in an indirect manner to other modifications

of structure. The early male progenitors of man were... probably furnished with great canine teeth; but as they gradually acquired the habit of using stones, clubs, or other weapons, for fighting with their enemies, they could have used their jaws and teeth less and less.... As the various mental faculties were gradually developed, the brain would almost certainly have become large.

In outline, Darwin hypothesized that free hands made it easier to use and craft tools—and that the escalating intellectual challenge of building better tools eventually led to a larger brain. It's an elegant story, but truth and elegance sometimes come apart. We now know that sophisticated tool use predates our species' expansion in cognitive capacity by a few million years. And if Lucy's story teaches us anything, then it's that the first thing our ancestors' hands were busy with were not tools but babies. Tool use did not make our brains bigger.

So, when did our heads get big?

Homo habilis did have a larger brain than any of the *Australopithecus* species, but the handyman and handywoman also had bigger bodies, so proportionally their brains weren't that much larger, if at all. A real jump in brain size happened in *Homo erectus*, but that took a while. The oldest *H. erectus* specimen had a brain between 600 and 900 cubic centimeters, but by about a million years ago, *H. erectus* brains were larger than 1,000 cubic centimeters. Remember that *H. erectus* was the first of our family to leave Africa and spread around the world, diversifying in response to different climatic conditions and giving rise to new species along the way.

It wasn't until our species came along that brain size really got out of whack. And that was "only" about 300,000 years ago. Because the brain is encapsulated in a skull, changes to the size of the brain, particularly significant changes, have an effect on the shape of the overall head and face. We'll get into the details of the link between brain size and skull shape in Chapter Four, but for now think about the ways your head differs from the head of other mammals.

Most other mammals have a long, narrow, tube-shaped head, but ours is wide, short, and round. We have a forehead and no snout. Our short and nearly vertical neck attaches to the spine in the center instead of at the back of the cranium. Our noses extend out from the face with downwardly oriented nostrils, which means if we want to locate a smell, we need to lift our head. And we have a chin below the lower jaw, for which no one has ever found a convincing function—and not due to a lack of trying. Then there are other features that may not be as obvious. We sweat profusely from the head, particularly from the face and scalp; our tongues are rounded and descend far into the neck. Oh, and our brains are more than five times larger than predicted by our body size.

But why did our brains get bigger?

To answer that question, it makes sense to start with *Homo erectus*. There is a reason *Homo erectus* could move out of Africa, effectively becoming the first hominin colonizer. They became the first colonizer, not because they walked out of Africa but because they ran. We humans may be pretty lousy sprinters compared to most quadrupeds, but we are pretty good long-distance runners, and we think *Homo erectus* was the first one. It's a guess, but an educated guess because they had a body quite like ours, bar our large brain. If you are a runner, you will know from experience that running makes you prone to falling forward, because the trunk and the neck have a tendency to pitch forward, particularly at heel strike. Someone once described running as perpetually trying not to fall flat on your face, and there is certainly some truth in that. To avoid such ignominy, both the head and the trunk need to be stabilized by powerful muscles—the back muscles, which connect to the neck muscles, and the muscles that form one of the most distinct features of humans, the greatly enlarged gluteus maximus, also known as your bum. To allow for the bigger muscles, the posterior iliac spine and the sacrum expanded in size to create a larger surface for the muscles to attach to.

Again, if you are a runner, you also know that it is impossible to run without bending your elbows and swinging your arms; particularly during the phase when both legs are off the ground, your arms and the rotation of the upper body are needed to keep your balance. We are much better at rotating our trunk relative to our hips than, say, are chimpanzees because of our elongated, narrow waist. Our wide shoulders allow us to use the arms as a counterbalance during running. At the same time, we no longer required massive forearm muscles, which were handy for climbing trees, but very expensive energetically. Much smaller forearm muscles (50 percent less massive relative to total body mass, compared to chimpanzees) make it easier to keep the elbows flexed during running. Lastly, having a flat face helps with balancing the head.

There are, of course, other changes—to leg muscles and tendons, to name a few—but the changes to the pelvis, shoulders, and forearms all left their traces in the fossil record. Note that these features have nothing to do with walking, which is different mechanically from running. And yes, that also means your bum is an adaptation for running, not sitting.

Before the invention of long-distance running, probably around the time *Homo* evolved, our ancestors had started to change their diet from predominantly feasting on ripe fruits to a more varied but less reliable diet. Because you are what you eat, you can get a good idea of the kind of food that was consumed by looking at the carbon isotopes in teeth. And fossil teeth we have aplenty. Carbon naturally occurs in different forms that differ in the number of neutrons in their nucleus. The stable carbon isotopes in carbon dioxide are used differently by different types of plants during photosynthesis. Plant eaters accumulate these stable carbon isotopes in the enamel of their teeth. That means we can analyze the proportion of naturally occurring carbon isotopes in the enamel of fossil teeth, so as to get an idea of what plants our ancestors ate. It's not quite as if we were sharing the dinner table with them, but it gives us a pretty good guess.

The isotope story tells us that our early ancestors began using plant foods that are rarely eaten by African apes today, probably because they were venturing farther afield now that they more easily moved around on the ground. Then came *Homo erectus*, the runner. Because of their newly found ability to move fast through their environment, *Homo erectus* could more effectively exploit yet another food source, one rich in proteins and fats, and that was meat. This addition to the diet significantly increased the amount of energy available, which in turn made it possible to sustain an organ known to be excessively costly: the brain.

An adult *Homo sapiens* needs to consume 2,800 to 4,200 calories of energy per day just to keep the brain running. To put that into perspective, a chimpanzee needs 1,000 to 1,200 calories per day. Given the average brain size of *Homo erectus* (let's make that 800 cubic centimeters), we can estimate they had to find about 2,000 calories per day just to maintain their brain. The addition of high-energy meat to the diet made it possible to sustain a larger, and thus more expensive, brain. Bigger brain, higher cognitive skills, great mover. For almost 2 million years, *Homo erectus* was the dominant hominin species, spreading around the world.

ON BABIES

Three newborns. The first immediately gets onto its legs and reaches out for mom's nipple, which it can only just reach. The second starts life with a six-foot fall, landing on its head, but nevertheless is up and running within an hour. The third takes months before it can even hold its head up. I think you know which one is the human newborn. (The first is an elephant calf, the second a giraffe.) Our babies are pathetic. And that is not because the mother doesn't put a lot of energy into them prior to birth.

Human mothers give birth to babies that are about 6 percent of their body mass. A chimpanzee's baby is about 3 percent of its mother's body mass even though both moms are more or less the same weight. Our babies are also exceptionally fat. So, we produce babies that are exceptionally large, yet exceptionally helpless. The two are connected. It is that hugely expensive organ again, the brain.

The conventional story, popularized by Stephen Jay Gould in his essay "Human Babies as Embryos," goes like this. Because the human pelvis became so narrow to allow more efficient bipedal locomotion, and because our brains got so big, human babies are born way earlier than they should be. In other words, they are underbaked. If our babies would be born at the "right" time, pregnancies would take about a year and a half—almost twice the normal time. The more scientific version of the story is called the "obstetrical dilemma."

Sadly, the idea of an obstetrical dilemma has been extrapolated, not by Gould but by others, to mean that women are compromised creatures because they cannot have the slim hips they need to be the best runners. And that is all evolution's fault. Women's hips cannot get narrower, which would make us better runners, as otherwise the babies couldn't get out. And so human females became the example of the ultimate compromise between the need to allow a safe passage for the baby and the ability to become an Olympic athlete. Human men, on the other hand, are perfectly adapted with their slender hips, unburdened by carrying the next generation. It is just that there is no evidence that women's skeletons are not as well adapted as those of males. In fact, there is so much variation among both men and women in the width of their hips that any claim women's hips constrain their ability to become fast runners is, to say the least, a little silly. But back to babies.

Are our babies special among primates or not? Of course, the answer is yes and no. Let's start with the no. Of all the primates, the great apes have the longest pregnancies, with a range among species of roughly 30 to 39 weeks, so our 38- to 40-week pregnancy does not seem that much of an outlier. Perhaps our babies are not that underbaked, time-wise that is. And when it comes to helplessness, chimpanzee babies are also pretty helpless for a long time. They rely completely on their mothers for nutrition, thermoregulation, and protection. In the first weeks they are even incapable of holding on to their mother without mom lending a helping hand. At the time of birth, a baby chimp's brain is about 40 percent of its future adult size. It certainly has a lot of growing and maturing to do, and mom will have to come up with the necessary resources. Not too different from our situation, is it?

Let's now get to the yes. At the time of birth, a human baby's brain is only 28 percent of its adult size, so when it comes to brain size, our babies are most definitely underbaked. It will take a further six to seven *years* before the child has a brain that is 95 percent that of its adult size, compared with three to four years in chimpanzees. And then there are the costs. At birth, the human brain accounts for approximately 87 percent of basal metabolic rate, which is the rate at which the body uses energy to stay alive. That drops to about 50 percent around its first birthday. Energetically, our babies are approximately 8.7 percent more costly than chimps for the first eighteen months. Now it probably also starts to make sense why our babies are the fattest of all. If you need to spend 87 percent of your energy on maintaining your brain, you don't want to risk not having enough energy to keep it running in case the milk supply is interrupted for a time. Having a lot of fat, then, is a great safeguard. But all that energy the baby accumulates before birth has to come from mom. And according to some, energetic demands are the real reason why human pregnancies are "only" nine months, which is way before our babies have a brain comparable to that of a chimpanzee.

Despite the immaturity of our newborns' brains, human mothers invest more in their baby prior to birth than one would expect of a mammal of our size; that increased investment translates into an additional thirty-seven days of pregnancy. Relatively, our newborn babies' brains and bodies (remember all that fat) are way bigger than those of the other great apes. It seems that at some stage in our evolutionary history we started to invest much more in our babies before they were even born. Yet, they still came out helpless. They had to, as mom could not continue to invest in her baby in utero. (And then there are those narrow hips.)

Gestation, the scientific name for pregnancy, is metabolically extremely expensive for the mother; not only does she need the energy to maintain her own body in pregnancy but she also needs to grow her baby. As the pregnancy progresses, the metabolic costs increase, until it is no longer possible for the mother to continue and the baby needs to leave her body. This scenario is true for all live-bearing animals, but in humans there is a special twist. And that twist is that the energetic constraints have everything to do with the large brain of the baby, which has driven the baby's need for fat reserves. While the original "obstetric dilemma" takes only the constraint of the female pelvis into consideration, the so-called energetics of gestation and growth (EGG) hypothesis places the fault for our babies' immaturity on their extremely expensive brain.

Regardless of where exactly one wants to place the emphasis, hips or energy, the outcome remains the same. Our babies are expensive before they are born and need a lot of help afterwards for a very long time. They need much more help and for much longer than any of our relatives' babies. You'd think we would produce them only sparingly. As I myself have shown, nothing could be further from the truth.

We produce babies at a prodigious rate, relatively speaking of course. Data from hunter-gatherer societies, in the absence of agriculture and modern medicine, shows that the so-called inter-birth interval in humans is 2.4 years, whereas in chimpanzees that number is 5.75 years. (At this stage, you may be wondering why I constantly compare humans to chimpanzees, given that chimpanzees are not our ancestors. It is the best we can do really; also, in many ways chimpanzees are rather like us. After all, we are closely related.) Age at weaning, when the youngster relies solely on solid food,* is also strikingly different between humans (2.4 years) and chimpanzees (5 years). That means that by the time the next child makes its appearance, the previous one has only just been weaned. And in humans, being weaned does not mean the child is independent. It seems we "ought" to wait many more years before we wean our children. And the next one shouldn't make its appearance that quickly, either.

If we look across mammals, we see that species with larger bodies and brains typically grow slower, their teeth erupt later, they produce their first offspring at a later age, their gestation period is longer, and so is their inter-birth interval. They also live longer than smaller mammals. We certainly fit the mold with respect to slow growth and longevity, but are completely off the scale when it comes to the other characteristics. Mothers in general are supposed to† start weaning their young when the

---

* Sometimes referred to as the end of weaning, with the beginning of weaning being the time when the first solids are introduced in the baby's diet.

† By "supposed to" I mean that natural selection has shaped their behavior so as to optimize their lifetime reproductive success.

metabolic requirements of the young exceed what the mother can provide through her milk. At that stage, the mother should add other foods besides milk to the young's diet until milk is no longer needed. Mammals other than humans typically wean their young when they have reached about one-third their adult body weight. If we extrapolate that rule of thumb to humans, then our children should be weaned completely when they are four to six years for girls and five to seven years for boys. And then there are teeth. You would expect that by the time a youngster starts to eat adult food, it would have its permanent teeth. Not if you are a human. Our first adult molars emerge between 4.7 and 7.1 years—long after we have started munching. And that means that after weaning we still need to be fed a special diet. It's a diet that the children are incapable of getting together because they do not have the motor and cognitive skills needed to know what can be eaten and how to prepare it. Combine that with exceptionally slow growth until the growth spurt around puberty kicks in, and you can see the utter dependence of our children.

And yet, we outcompeted all other hominin species, dominate the earth, and are rapidly destroying the planet's natural environment. How?

HIPS BEFORE BRAINS

Darwin was right. It all started with walking upright. But as we've seen, it's not quite the way he envisioned. Perhaps he didn't get everything right—and who does?—but he certainly had some amazing insights into our evolutionary history, as the following from *The Descent of Man* illustrates:

> With strictly social animals, natural selection sometimes acts indirectly on the individual through the preservation of variations which are beneficial only to the community.... With the higher animals, I am not aware that any structure has been modified solely for the good of the community.... In regard to certain mental faculties [of man] the case ... is wholly different; for these faculties have been chiefly, or even exclusively, gained for the benefit of the community; the individuals composing the community being at the same time indirectly benefited.

What Darwin hinted at was the idea that we evolved our mental capacities not because they do us any good personally but because they benefit the community. The community being our family and other relatives. I think he was spot on.

Ever since our early ancestors decided to leave the trees, their bodies adapted to their erect posture. They moved into different environments, and learned to eat different foods; some of those foods contained more nutrients, allowing their brain to become larger. But as with all changes, there are benefits and downsides. Females now had to carry their babies, making those mothers even more vulnerable to predators and nasty neighbors. And the babies probably needed some help being born in the first place, as they had to wiggle their shoulders through a skinny birth canal. Sticking together—mothers, fathers, siblings, cousins, and others—seemed like the way to go.

Then, through a fluke of nature, the brain of *Homo sapiens* ballooned, leading to arguably the most expensive offspring ever to be produced. They were offspring so costly that mothers stood no chance of raising their babies alone. Mothers and their babies, both before and after birth, still need all the help they can get. Their big brains are both the cause and the solution to the problem, as Darwin said, "No one, I presume, doubts that the large size of the brain in man, relatively to his body . . . is closely connected with his higher mental powers." We used our increased mental powers to create communities to foster and raise our children—children who had become utterly dependent on us because of their large brains.

Having summarized the road to the childcare problem with which this book begun, and having hinted at its solution, it is time to sift through the scientific evidence for how our skeletons changed to allow us to become bipedal and what exactly made our brain grow.

In the beginning . . . our embryonic development slowed down.

three

BEAT OF A DIFFERENT DRUM

Unlike everyone else, by the time he turned fifty, Mickey Mouse looked much younger than when he made his debut in *Steamboat Willie* in 1928. Not only did he look much younger at fifty but he also had become much nicer. The original Mickey Mouse was a rambunctious and even slightly sadistic little fellow, who tried to kill a parrot twice because it laughed at him, and used a goat, a cat, a duck, piglets, and a cow as musical instruments. The goat had just eaten Minnie's violin, so Mickey and Minnie used the goat as a phonograph, Minnie "cranking" the goat's tail to play "Turkey in the Straw," the sheet music for which the goat had also just eaten. While Minnie cranked the goat, Mickey pulled on the cat's tail and swung it above his head to make it scream; he used the duck as a makeshift bagpipe, pulled on the piglets' tails to make them squeak, and used the teeth of the cow as a xylophone. Mickey's mischievous streak didn't last long, though; as he became a national symbol, his appearance slowly changed until he had morphed into the well-behaved and lovable character he is today.

As Mickey Mouse's character softened, he started to look more childlike. His arms and legs became thicker, his head larger, his snout less protruding, and his eyes bigger. Mickey's ears moved back, so the distance between his nose and ears increased, and his head became more rounded with a sloping forehead. Mickey Mouse had become older in age, yet younger in appearance. Before you think I have gone insane, writing about a cartoon character in a book on human evolution, know that Stephen Jay Gould used Mickey Mouse's rejuvenation to explain an important process in evolution. That evolutionary process is neoteny, the progressive juvenilization over evolutionary time caused by the delay in or slowing down of the physiological development of an organism.

Probably the most famous example of a neotenic animal is the axolotl, *Ambystoma mexicanum*, a kind of salamander. In the normal course of events, a salamander's gills gradually become lungs, so the animal can live on land. Not so with the axolotl, which retains its external gills and accordingly spends its whole life underwater. The species has gotten "stuck" in a juvenile developmental stage, but its legs still grow, so it's colloquially known as the "walking fish." Crucially, with neoteny, sexual development *isn't* held back. So, what you get is a sexually mature adult that looks juvenile relative to its immediate ancestor.

Some organisms take neoteny to the extreme. In some fly species, the larvae don't bother turning into adults before producing their own offspring, so small larvae crawl out of larger larvae. Because there is no sex involved, all these larvae produced by the older larvae are females. After a few such rather extraordinary generations, the larvae pupate and turn into males and females who mate and start the bizarre cycle again. Surely material for a future sci-fi horror movie.

Mickey Mouse's transition from a ratty appearance at his creation to a more childlike and positively cute character in later years illustrates, in one fictitious lifetime, our own evolutionary history. We humans retain the juvenile features of our ancestors into adulthood. Our ancient forebearers, like Mickey Mouse in *Steamboat Willie*, had protruding jaws, small heads, and vaulted craniums. Using Gould's own words, "our evolutionary history is characterized by a general, temporal retardation of development." It's evolution by retardation. We humans are retarded in the development of our skull, as it doesn't go through the later stages of development. So, our head remains round and bulbous and the bottom of our skull, the cranial base, looks comparatively stunted. The result is that we look more like a baby chimp than an adult chimp.

Imagine a very young vertebrate embryo as a comma, where the thickest part of the comma will form the head. As the embryo develops, the comma "unfolds" so that during later development, the animal's head points in a direction that is the continuation of the line of the backbone. To connect the head, the spine passes through an aperture in the base of the skull called the *foramen magnum*. As our tree-swinging cousins develop in utero, this opening migrates to the back of their heads. But in our case, the foramen magnum stays centrally positioned, so our head sits squarely atop the spine.

Now, imagine looking at a picture of yourself in profile. Draw an imaginary line from your eyebrow across the top of your ear, and then another line down from your ear to the base of your skull. The angle at your ear that is formed by the intersection of these two lines is about 100 degrees—much closer to a right angle than a straight line—and is known as the *cranial base inflexion*. During the unfolding of the comma, if the cranial base inflexion decreases—that is, the angle of the two lines increases—it pushes out the developing jaws, forming a snout in animals that have a snout. In primates, though, the early embryonic curvature is retained, so the head points in a direction that is at a right angle to the axis of the body and no snout is formed. Both humans and chimpanzees are born with a flat face, but as chimpanzees grow into adulthood, the cranial base begins to flatten (increasing the angle of the two lines just described), basically "pushing" out the jaws so that adult chimpanzees do end up having a snout.

Perhaps it is easiest to see the developing head as a round squishy ball. Then imagine someone pressing down on the ball while holding three sides firmly in place. The downward pressure will then push out the head in the only direction it is free to move, forming a protrusion, which is the snout. In human development, no one places pressure on our head, so we never form a snout.

The change in the position of the foramen magnum started a long time ago, probably as early as in our old relative *Sahelanthropus tchadensis*. Just by looking at its position, paleoanthropologists could, and still can, determine if the skull belonged to a bipedal species. Foramen magnum central? Bipedal. Positioned toward the back of the skull? Quadrupedal. A centrally located foramen magnum gave our ancestors no choice: They had to start walking upright. If they would have walked on all fours, they'd be looking straight down; or they would be uncomfortably craning their necks upward for a view of what's ahead. A video of Pedals shows how uneasy he must have been, holding his head in an awkward position while standing upright.

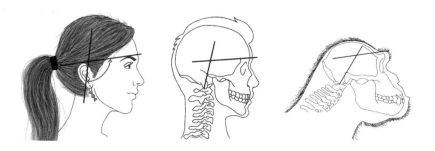

Pedals had to make changes to his skeleton *after* his skeleton was developed, and so his options for change were limited. His embryonic skeleton, on the other hand, would have been much more malleable.

INFLUENTIAL FAMILIES

Jean Louis Rodolphe Agassiz was an extremely influential but seriously flawed man. The long-lasting legacy of the Swiss-born American biologist and zoologist is the Museum of Comparative Zoology, established in 1859, at Harvard University. Instead of a randomly displayed collection of view-worthy organisms, Agassiz wanted his museum specimens to be arranged so they explained some theory of nature. Sadly, the theory he selected was the wrong one. He believed that God had created all living organisms and had placed them at specific locations around the earth, where they had remained ever since. In creating all living beings, God had also made some organisms more complex than others, leading to a hierarchy—a hierarchy that scientists needed to figure out. (If that sounds like Aristotle's idea, it is probably no coincidence.) It was this hierarchy that was displayed in Agassiz's collection of comparative zoology.

Louis Agassiz was not convinced by Charles Darwin's theory. He believed Darwin's theory was "a scientific mistake, untrue in its facts, unscientific in its methods, and mischievous in its tendency." Yet Agassiz had the evidence for the common descent of organisms literally in his hand—if only he hadn't been blinkered, and so dismissive of Darwin's theory, to notice. An embryo of some sort of vertebrate animal he was studying had no label, and Agassiz could not tell if the embryo was from a mammal, a bird, or a reptile. Darwin used this anecdote to illustrate how studying the development of embryos reveals evolutionary relationships.

The similarities in the embryonic stages of very different animals had long puzzled naturalists. Why is it that animals that look nothing alike as adults are remarkably similar as embryos, especially as very young embryos? One idea that long held sway was that of *recapitulation*—the notion that organisms recapitulate, or run through, primitive stages as they develop. Recapitulation is not a form of neoteny, because development is not halted at an early stage. Instead, with recapitulation, a "higher" animal goes through the embryonic stages of "lower" animals before turning into its unique adult form. "Lower" animals simply remain "stuck" in a more primitive embryonic form. Or, so goes the obsolete idea. "Higher" and "lower," of course, referred to the position of the animals in Aristotle's *scala naturae*. Before publishing *On the Origin of Species*, Darwin became an expert on the study of barnacles. He had two reasons for starting this "side project," a little digression that in the end took him eight years and led to a four-volume book. Darwin felt it necessary to become an expert in one group of organisms before publishing his grand theory. Otherwise, he feared, he ran the risk of not being taken seriously and his life's work too easily dismissed. Additionally, he wanted to finally solve the mystery of Mr. Anthrobalanus, a strange barnacle Darwin found in a hole in a conch shell that had been collected from a beach in Chile during his HMS *Beagle* voyage. Barnacles normally make their own little shelters, so Mr. Anthrobalanus seemed an odd specimen indeed. On closer inspection, Mr. Anthrobalanus was even more odd because it had joints, unlike other barnacles.

At the time it was generally believed that barnacles were a kind of mollusk, simply because many adult barnacles look a great deal like, say, limpets. But Darwin was not fooled by superficial similarities, and he took a close look at the young barnacles, the larvae. Unlike Agassiz, Darwin knew that young life stages—larvae or embryos—can look eerily similar, not because they go through the same "lower" developmental stages but because they are closely related. By looking at the barnacle larvae, Darwin could tell that barnacles are not a kind of limpet but, rather, a relative of shrimps, lobsters, and crayfish. And Mr. Anthrobalanus? He remains the odd one out. Thanks to modern molecular techniques, we now know that Darwin's odd barnacle belongs in a special group.

Ernst Haeckel, the German zoologist who inspired the young Marie Eugène Dubois, took Darwin's idea of descent with modification and combined it with the older idea of recapitulation to come up with his 1866 biogenetic law, better known by the catch phrase *ontogeny recapitulates phylogeny*. According to Haeckel's biogenetic law, the stages an animal embryo undergoes during development are a chronological replay of that species' past evolutionary forms. The biogenetic law explains why all vertebrate embryos, at some stage in their development, have a series of bony "arches" that support the gills in fish (gill arches)—that is, they all have a fish ancestor.

Haeckel is famous for his exquisite drawings of animal forms, but some of his drawings also got him into trouble. Drawing embryos from a fish, salamander, tortoise, chicken, pig, cow, rabbit, and human, Haeckel wanted to convey the similarities among the different vertebrate groups during their embryonic development. Apparently, Haeckel had substituted a dog embryo for the human embryo, maybe because he did not have access to a human embryo. Because of their similarity, though, Haeckel's fraud does not detract from the point he wanted to make. But while the similarities are real, Haeckel's biogenetic law is not real. Vertebrate embryos do not have gill arches because nature requires them to recapitulate their phylogenetic history.

The bony structures that in fish will become the support for the gills, the gill arches, serve a different purpose in other vertebrates. We had gill arches, not because we needed them to get oxygen while floating in fluid but because we—vertebrates—all use the same family of genes to construct our necks. Mother Nature's ability to repurpose genetic programs via the *timing* of gene expression is one reason organisms can be genetically very similar yet morphologically quite different. Not wanting to waste innovations, natural selection is good at putting old wine into new bottles.

It's time for another cartoon character, this one Sonic the Hedgehog. Sonic the Hedgehog is the hero in a video-game series of the same name, who battles the evil Doctor Eggman, a mad scientist who wants to conquer the world (don't we all?).

The cartoon character Sonic the Hedgehog can run at supersonic speeds and curl into a ball to attack his enemies. In the academic biology game, a gene named *SHH* codes for a protein known as Sonic Hedgehog* (Shh). This Sonic Hedgehog protein, or Shh, functions as a chemical signal essential for embryonic development. Sonic Hedgehog is a so-called *morphogen*, a substance whose nonuniform distribution affects the pattern of tissue development during embryogenesis. The concentration of Shh within a cell determines what kind of cell is produced through interactions with other gene products, thereby guiding the development of an embryo from a single cell to an organism composed of many different cells and cell types. Different concentrations of Shh carry different instructions, so a single gene can fulfill different functions depending on where exactly it is located and in what concentration its protein products are present. One of Sonic Hedgehog's functions is the development of the gill arches in vertebrates. (If you are puzzled about the analogy of the cartoon Sonic the Hedgehog and the protein, then you are not alone. I don't get it, either.)

SHH is a member of the hedgehog gene family. The hedgehog family was first discovered in the fruit fly, but is now known in almost all animals and even in protists (organisms that are neither plant, nor animal, nor fungus; in other words, protists are a sort of leftover group). The hedgehogs are an old family, probably stemming from the beginning of the evolution of all eukaryotes (all organisms except bacteria and viruses). One can trace back the evolutionary history of the gene family by comparing

---

* Gene names are italicized, but their protein products are not.

the proteins produced by the genes in the family in different taxa. Recall that just as one can read DNA by determining the sequence of nucleotides, one can read a protein sequence by looking at the sequence of amino acids. One can then determine where the gene family originated in the tree of life (or here, the eukaryotic tree) and see how particular genes have been adapted for a specific purpose by natural selection in separate taxa.

The expression of *SHH* is regulated by yet another gene family, the *Hox* gene family. *Hox* genes code for transcription factors—proteins that modulate the expression of other genes. Found in all animals, the *Hox* gene family stands out because the order in which the genes are found in the genome means something. While all other genes are found in random locations in an organism's genome, *Hox* genes are organized along a chromosome in exactly the order in which they are expressed in the embryo. Anterior genes are expressed earlier and, as you might guess, at the front of the embryo, while posterior genes are switched on later in development and in more distal parts of the embryo's body. This pattern of expression is the same in all animals, from the humble fly to an elephant, to a human.

The more complex the organism's body plan, the more *Hox* genes it possesses. During the evolution of vertebrates, a single ancestral cluster of 13 *Hox* genes probably duplicated twice to form a complete set of 52 *Hox* genes in four clusters—four copies of each of the 13 *Hox* genes. The fly, like all invertebrates, has only a single cluster of 13 *Hox* genes. Mice, elephants, humans, and all other organisms with four limbs (tetrapods) have 39 Hox genes distributed over four clusters. At some point in evolutionary history, some of the individual *Hox* genes got lost, as no extant organism (organisms still present today) possesses the full set of 52 ancestral *Hox* genes. The timing of the expression of *Hox* genes affects the different body plans we see in the animal kingdom. That timing relies on the *segmentation clock*. (We are finally getting to the clock that retarded our development and kicked off the changes that allowed us to walk upright.)

The segmentation clock is made of genes and proteins, and it keeps the time inside the cells. Such timekeepers make sure that the right genes are turned on and off at the correct time. The segmentation clock regulates the creation of repeating body segments in the developing embryo and paired structures such as limbs, in concert with the *Hox* genes. The segmentation clock's main instrument is a gene with the exciting name of hes family bHLH transcription factor 7, or *Hes7* for short. *Hes7* is a mammalian gene; other taxa have a different gene that performs the same function (birds have *hairy1*, illustrating how the naming of genes is an art in itself; ever seen a hairy bird?). There is more than one gene involved in the segmentation clock, but *Hes7* is the gene that gets it all going—it is the key pacemaker.

An animal embryo grows and develops by the addition of segments, called *somites*. In vertebrates, somites become vertebrae, for example. The timing of somite formation determines how large an organism becomes. Giraffes do not have a longer neck because they have more vertebrae than a mouse but, rather, because their vertebrae grow over a longer period, so the giraffe's neck becomes longer. Very young embryos all develop at similar speeds, explaining the similarities that Haeckel portrayed. But soon they each go their own species-specific way under the influence of *Hes7*.

The activity, or expression, of *Hes7* oscillates. When the gene is active, it produces the protein Hes7, which builds up until there is so much *Hes7* that the gene is deactivated and is switched off. Once all the protein is broken down, the gene is switched on again, and *Hes7* is produced again, starting the cycle anew. It's a nice example of self-regulation, as it is the gene's product itself that silences the gene once there is enough protein. *Hes7* is the pacemaker of the Notch signaling pathway—a pathway that regulates gene expression of, among others, *Hox* genes. I see embryonic development as a game of dominoes—the version in which one constructs a chain reaction. If they are set up correctly, you need push only the first domino for the whole construction to fall over, one domino at a time. *Hes7* initiates the Notch pathway, which activates the *Hox* genes, which segment and grow the embryo. *Hes7* is the first domino to fall.

The timing of the *Hes7* oscillations differ among taxa. In human embryos, each oscillation takes five to six hours. In mice, it takes two to three hours. The difference in oscillation timing is due to the different speeds at which the Hes7 protein degrades in humans and mice. In humans, the protein hangs around for much longer, so it takes longer for the next segment to form. No one knows what causes the different rates of protein breakdown among the taxa, but it is definitively a relatively simple mechanism to use the same gene and adapt it to the species it is in. Subtle changes in gene-regulatory networks—networks of genes and their products that interact with each other to determine the function of the cell—are at the heart of embryonic development. No novel genes are required—just a little bit of tinkering does the trick.

Because of the striking similarities in the limb structures of vertebrates, Darwin realized they all shared a common ancestor. He wrote, "What can be more curious than that the hand of a man, formed for grasping, that of a mole for digging, the leg of a horse, the paddle of a porpoise, and the wing of a bat, should all include similar bones, in the same relative positions?" All these bones are in the same relative positions because they are all the product of similar sets of genes, whose expression is regulated by similar regulatory networks. Even though Darwin had no idea about the underlying mechanisms, he was correct when he ended *The Descent of Man* with the following statement: "Man still bears in his bodily frame the indelible stamp of his lowly origin."

Stephen Jay Gould, too, was a visionary. Without knowing anything about the molecular mechanisms, he envisioned that the growth of the skull is regulated by a "clock mechanism" that is running slower in humans relative to our ancestors and closest relatives. Because of the slow-running clock, our head reaches its shape and size at a much later developmental age than our ancestors' heads. Our sluggish clock slowed down our development, thereby providing us with unprecedented benefits, such as a hole in the skull ideally positioned to balance the head upright, as well as smaller jaws and teeth and, ultimately, space for a larger brain.

Okay, it's easy to imagine how a little bit of evolutionary tinkering modified the speed of a molecular clock, slowing down our embryonic development and ultimately changing our head so we are better off walking on two legs. But what is the evidence that having a centrally positioned foramen magnum helps an upright position? Let's take a quick dip into comparative analysis.

If the position of the foramen magnum really matters with respect to the way we move about, we would expect that bipedal species have a more centrally located foramen magnum compared to their quadrupedal relatives. We can also look at animals that spend a lot of time sitting upright (called "orthograde trunk posture") because they, too, would benefit from having a head balanced on top of their spine. There are plenty of species available for what evolutionary biologists call "comparative analysis." For example, we can compare bipedal kangaroos and wallabies to quadrupedal marsupials, bipedal kangaroo rats and jerboas (hopping desert rodents) to quadrupedal rodents. And the group of primates known as the Strepsirrhini, are perfect to compare upright-sitting species to related species that do not habitually sit upright. This last group includes the lemurs of Madagascar, galagos and pottos from Africa, and the lorises from India and Southeast Asia, and it includes species with and without an orthograde trunk posture. The Strepsirrhini are interesting for another reason, too; the group includes the only known venomous primate—the slow loris, native to Southeast Asia. When threatened, the slow loris licks its sweat gland on its arm to excrete the precursor of a toxin, which when mixed with saliva becomes truly toxic. It then bites its enemy with toxin-coated teeth. A slow loris's enemy is mostly another slow loris. Instead of biting each other, slow lorises would be better off biting the people who capture them and pull out their teeth before selling them. (Slow lorises are popular exotic pets because their large eyes and small ears make them adorable.)

As one might predict, the location of the hole in the skull is associated with the way a species moves around or sits about. Bipedal marsupials and rodents have a foramen magnum that is positioned more toward the front of the skull than it is in related marsupials and rodents that are quadrupedal. Strepsirrhini that spend a lot of time upright also have a more forward foramen magnum relative to those that do not sit upright but, rather, hang down from tree branches. And then, of course, there are our own relatives—chimpanzees, gorillas, and orangutans. Our foramen magnum is so much more forward than it is in chimpanzees, gorillas, and orangutans that it's possible to tell from position alone that it is a human skull.

But holding our head upright was only the beginning.

SMALL CHANGE, BIG EFFECT

If you ever had a pet mouse from the Lathrop mouse farm in Massachusetts, you have seen the effect of a single mutation in one of the eight known bone morphogenetic protein (*BMP*) genes. This mutation has interesting morphological effects that are loved, apparently, by certain mouse aficionados. Mice that carry a *BMP* mutation have very short ears, a wider skull, and a shorter nose than normal mice. They look cute. And cute is exactly how their original breeder, Abbie E. C. Lathrop, wanted them to look. In a twist of luck, the cute-looking mutated mice have also helped elucidate the molecular signals that control the formation of bones and cartilage.

Bone morphogenetic proteins are signaling molecules responsible for the formation of bones and joints during normal development. Signaling molecules tell the target cell when a particular gene should be expressed, meaning the specific sequence of DNA is read and translated into a protein. Because they specifically target a cell, signaling molecules act locally—say, in a distinct part of the developing embryo or at a particular time during development. Increasing the activity of a *BMP* in a new body region—a region that is the result of an evolutionary innovation, for example—will kick off the formation of a new skeletal element. Similarly, a decrease in the activity of a particular *BMP* will reduce the size or alter the shape of skeletal elements that normally depend on that *BMP*. Lathrop mice lack one *BMP*, leading to mice that look like a different species altogether: voles. Voles belong to the family Microtidae, which means "small ears." A single mutation can thus change one species into another—at least as far as this aspect of their morphology is concerned.

Other mice also show us how small genetic changes can have significant morphological effects. Unlike the Lathrop mice, though, these mice did not end up living a happy life as someone's pet. I am talking about transgenic mice—mice whose genomes have been manipulated to either add or remove a particular DNA sequence when they are an embryo. Because the studies I describe here had the aim of finding out how genetic changes affect the embryonic development of specific body parts, these transgenic mice were never carried to term. But first, a new term: *enhancer*. And a major distraction.

Enhancers control cell-type specific gene expression via the production of transcription factors—proteins that control the rate at which information in the DNA is transcribed into an intermediate molecule, called messenger RNA, by binding to a specific DNA sequence. When a gene is transcribed, the double helix opens at the location of the gene, and the information encoded in the sequence of nucleotides is transcribed, or copied, into a single strand of messenger RNA. But there are instances where, without a specific enhancer, a particular gene is not expressed at all. And that brings us to the distraction.

Males of many mammal species—and that includes our chimpanzee cousins—have penile spines, or keratinized structures along the glans or shaft of the penis. It is not always clear what their function is, but in cats these spines seem to trigger ovulation in the female the tomcat mates with, while in other species the penile spines may cause the formation of a "genital lock" during mating, so the female cannot quickly mate again. Humans are the only primates so far found to lack a specific enhancer—the so-called androgen receptor (AR) enhancer—needed for tissues to respond to circulating androgens such as testosterone. Because the AR enhancer got lost in our evolutionary past, human male penises are spineless. Human males are also whisker-less, as the same enhancer causes the growth of facial whiskers. Why?

The loss of whiskers is probably a side effect of the loss of the AR receptor, as it is difficult to come up with a good reason why having whiskers may be bad. The lack of penile spines seems easier to explain, although admittedly the explanation remains speculative. It has to do with pair bonding. A less prickly penis makes mating a more enjoyable event for the female so she is more likely to mate again with that same male, or so goes the argument. In mammals, there is a strong association between males having less armored penises and strong monogamous pair bonds. That makes sense if females are more likely to stick with males when mating is more pleasant. And so it seems that human males have lost their spikes because pair bonding is so important. But why is pair bonding so important in humans? To ensure the survival of our needy babies, whose needy-ness this book has set out to explain. So, perhaps this wasn't such a distraction from the main story, and instead is a small piece of the puzzle.

Back to enhancers—specifically their role in organizing the formation and molding of bones and cartilage. We humans lack another enhancer, this time one that affects the expression of one particular bone morphogenetic protein gene: growth differentiation factor 6 (*GDF6*). And here is where the transgenic mice come into the picture. By removing the enhancer from mice, we can figure out what the consequences are of not having that enhancer, and by extension, what the enhancer is supposed to be good for.

Remember, we are trying to figure out what changes were needed to allow our ancestors to leave the trees and become bipedal. Mice embryos that, like humans, lacked the *GDF6* enhancer developed just like mice tend to do, with one exception. Their feet became different. The toes became shorter, the big toe was no longer opposable, and a particular foot muscle—the abductor hallicus muscle—became smaller. Feet with long toes and an opposable big toe, plus a large abductor hallicus muscle, are adaptations for tree climbing. So, one small genetic change—the loss of the enhancer influencing the expression of *GDF6*—turns a chimpanzee-like foot into a human-like foot. But arms and hands needed to change, too, as no longer required for hanging off tree branches. Another small change with large consequences.

Another study, also using transgenic mice, provided the insight. Here, the change did not result in a loss, as in the whisker and spines examples, but in a so-called gain of function. The function gained in this case was a new pattern of gene expression, which as it turned out plays a role in the formation of important structures—the hands and arms. The stretch of DNA in question carries the name human-accelerated conserved noncoding sequence 1, or *HACNS1*. Extremely similar versions of *HACNS1* are found in nearly all vertebrates sequenced so far, hence the term "conserved." Not so in humans. In humans, the sequence seems to have changed drastically over evolutionary time. Because *HASNC1* is so conserved in all other species, it seems likely that natural selection has tinkered with it to serve a human-specific purpose.

And indeed! Introducing the human version of *HASNC1* into mice led to a change in gene-expression patterns in the forelimbs of the mice, particularly in the mice's "hands," wrists, and shoulders. The mice that carried the human version of *HACNS1* did not actually develop human-like arms and hands (they were never born) but the experimenters used a clever trick to visualize gene expression in embryonic tissue. Only mice embryos that carried the human version of *HACNS1* showed gene expression where the hand, wrist, and shoulder would form. That means the human version of *HACNS1* functions as an enhancer, influencing the expression of genes that, in turn, organize the formation of arms. The change in gene expression could be traced to only thirteen base-pair changes—a tiny fraction of the genome.

These are small genetic changes with large consequences for how bodies come to look and what they can do. Subtle changes to the running of a molecular clock, a single mutation to a bone-growing gene, the disappearance of or change to enhancers. The resulting morphological effects provided our ancestors with an advantage, allowing them to start exploring new environments, which led, ultimately, to the diversification we now see reflected in the fossil record.

All that because of a weird mutation.

FROM 24 TO 23

Oops, we lost a chromosome! Through a fluke of nature, another weird mutation: two chromosomes got stuck together and remained so ever since. While gorillas and chimpanzees have 24 pairs of chromosomes, humans have only 23 (hence, the 23andMe DNA testing kits). Our chromosome 2 was formed by the head-to-head fusion of two ancestral chromosomes that remained separate in chimpanzees and gorillas and in all other primates. Because we are the only primates to have 23 pairs of chromosomes, the most parsimonious scenario is that the fluke of nature kicked off the separation of the lineage that would ultimately become chimpanzees and bonobos—and the lineage that includes us. That is, the origin of our species probably came about through this chromosomal fusion. Geneticists have long known that flukes like the loss of a chromosome, or in this case a massive reorganization, often lead to speciation events. To see why, we need to talk about sex.

Sex in biology is shorthand for "sexual reproduction," the process by which two parents combine their genetic material to produce another individual, or individuals—their offspring. Sounds simple, but even before parents get together, a lot needs to happen. First, they have to make sex cells.

Imagine what would happen if each parent would simply pass on their chromosomes to their offspring. Then with each generation, the number of chromosomes an individual carries would double. To avoid such an infinite increase in genetic material, each sex cell, or gamete, gets only one copy of every chromosome pair. And for that to happen, the homologous chromosomes—those that carry the same genes—first need to find each other and pair up so they can then be divided equally among the gametes, eggs, or sperm.

The chromosomes are first copied so there are two pairs of homologous chromosomes. Then, with each sexual cell division, called *meiosis*, four daughter cells are produced, each carrying only one of the homologous chromosomes. Now, each daughter cell carries only half the original genetic material. Males produce four sperm cells, but in females, three of the four cells never develop, so only a single egg cell is made.

Given how complicated the process is, it perhaps isn't surprising that a lot can go wrong when making these sex cells. Sometimes the gametes don't receive a copy of every chromosome, or they get too many, or just bits and pieces of some. In fact, most early miscarriages (within the first trimester) in humans are due to chromosomal aberrations. When something went wrong in our early ancestor and two chromosomes became one, no DNA got lost; it was just rearranged. But that doesn't mean the fusion did not cause problems, as now it was likely that two individuals would try and mate that did not have the same number of chromosomes.

Imagine that an individual with 24 pairs of chromosomes mates with another individual who has only 23 pairs. The first parent will make gametes with 24 chromosomes, the other parent will produce gametes with 23 chromosomes. Oblivious to its unusual genetics, the offspring probably will happily go about its life because it will still carry two copies of each gene variant, called *alleles*. But the trouble starts when it is time to produce gametes. Lining up chromosome pairs to ensure an equal distribution to each gamete is not so easy if you carry both the "new" fused chromosome 2 *and* the two "old" chromosomes. It leads to a mixed bag of gametes, with some, by chance, receiving the complete single set of chromosomes, either 23 or 24. But most gametes will be pretty mixed up, containing the "new" fused chromosome plus only one of the original chromosomes. Or, none of the new and old chromosomes, or all of them. Whatever the exact configuration, the point is that a hybrid of a 23-chromosomal individual and a 24-chromosomal individual would probably have very low fertility, with only a few of its gametes being in order.

Or, perhaps it will have none at all, making it completely sterile, like a mule. A mule is the offspring of a donkey father and a horse mother—two different species. A horse has 32 pairs of chromosomes and a donkey has 31, so a mule gets 32 chromosomes from the mother horse and 31 from the father donkey. That just doesn't work when the mule makes gametes, because mule gametes are a combination of horse chromosomes and donkey chromosomes. For some reason, the odd combination works in the mule, but it doesn't go any further. (An additional reason for mule offspring to be nonviable is the species' specific marks present on the chromosomes. Because mules have chromosomes of two different species, and therefore their sex cells do, too, things get pretty messed up.)

You'd think that individuals with 23 pairs of chromosomes would quickly disappear, given their reproductive problems when they mate with individuals with 24 pairs, which constitute the majority. Well, that didn't happen, did it? I can think of two, nonmutually exclusive scenarios that helped our ancestors get over their fertility problem.

Scenario #1: The First-Ever Version of Tinder?

Despite some stiff competition, the dating app Tinder wins hands down in its app category. On a single day (March 29, 2020), Tinder registered 3 billion swipes globally. That is a lot of people looking for the perfect partner. Tinder's aim is to help people find that perfect partner by matching their profiles using artificial intelligence. Each time you swipe someone to the left, that person's personal "Elo score" goes down. The *Elo score* is a term the chess world uses to rank the skill level of players; Tinder uses the score to rank people's "desirability." The more your profile is swiped to the left, the lower your Elo score becomes. The more people swipe to the right when your profile shows up, the higher your desirability.

The idea behind Tinder's success is to match people with the same level of desirability, so you would see only the profiles of people with similar scores as your own. But who decides on that first score? After all, you need to be part of the system before the algorithm can start updating your desirability based on what other Tinder users think of you. That decision, apparently, is made by Tinder's executive team, undoubtedly all individuals with above-average desirability scores. Guess what the initial score is based on? Looks.

One could object to Tinder's assumption that attractive people prefer to partner with equally attractive people, but the reality is that, in general, in the biological world, "like prefers like." There is a whole suite of reasons why that might be so, but probably the most important is that it avoids wasting time and effort mating with someone who does not belong to the same species. While hybridization can lead to viable offspring, as the traces of Neanderthals and Denisovans in our DNA have shown, hybridization between species almost always results in lower fertility. Best to avoid hybridization altogether, then. That choice could potentially be based on looks.

Like a good detective, I am going to speculate a little. Actually, I am going to speculate quite a bit. Imagine that the individuals in which the two chromosomes fused to form one chromosome looked different from those who still had the original 24 pairs of chromosomes. If that were the case, hybrids from parents carrying different numbers of chromosomes could then simply be avoided. Sounds unlikely? Perhaps, or perhaps not. We know from many plant species that the number of chromosomes a plant carries has a huge effect on the way the plant looks. Changes in the number of chromosomes can affect the number of flowers per plant or the size of the plant. Generally, the more chromosome copies, the larger the plant, a phenomenon that has been exploited by plant breeders for the simple reason that bigger plants yield more.

In humans, too, we see that a change in chromosome number can lead to big changes in the way a person looks. Having three copies of chromosome 21 (the cause of Down syndrome) or part thereof is associated with a flattened face, especially the bridge of the nose, almond-shaped eyes that slant up, and a short neck. Males who carry a Y-chromosome plus two X-chromosomes instead of one (Klinefelter syndrome) tend to be much taller than men with a single X-chromosome and are less hairy. Women who have only one X-chromosome (Turner syndrome) have a short neck, a low hairline at the back of the head, ears that sit low, and very narrow fingernails and toenails.

In all these examples, the phenotypic effect of the change in chromosome number is due to an increase in gene expression levels, simply because there are more copies of (some of) the same genes. Or it can be the reverse, when a chromosome is missing, as in women with only one copy of the X-chromosome, so there is not enough expression of particular genes. The situation in our ancestors was clearly different, as they did not lose or gain any genes but, rather, just moved them around when two chromosomes combined into one. I still like to think that there were some perhaps only subtle differences in looks between those with 23 and 24 pairs of chromosomes. Enough differences lead to positive assortative mating between individuals with the same number of chromosomes. Like chooses like. So, our ancestors became an *incipient species,* a group of organisms about to become genetically isolated from the last common ancestor we shared with chimpanzees.

Scenario #2: Be Like a Cane Toad

In 1935, the Australian Bureau of Sugar Experiment Stations introduced 102 cane toads from Hawaii into sugarcane country in Far North Queensland, where I now live. The cane toads were originally from South America, but had previously been introduced in Hawaii. The first arrivals in Australia were kept in captivity and nurtured to produce toadlets. By 1937, the initial cane toads had produced about 62,000 little toadlets, and these were then released into the abundant sugarcane fields of Australia's tropical north. The idea was that the cane toads would eat cane beetles, which are native to Australia and were devastating the sugarcane, an introduced crop. But, alas, the toads did not stick to the sugarcane fields and soon started to wander off. They have now spread across more than half the continent, even traversing areas they were never expected to be able to cross, such as arid environments and even deserts. It is estimated that there are more than 200 million toads currently hopping around in Australia.

There are a host of reasons why the toads are such successful dispersers in Australia, but I want to focus on their legs. Cane toads that have longer legs can, not surprisingly, move faster. As the toads spread from their initial point of release, those with longer legs left behind the ones with shorter legs. Imagine their dispersal through the country as the concentric ripples caused by a stone dropped in the middle of a lake. As they move farther out, the rings become more and more spatially separated. Similarly, the toads in the outermost ring bump only into other toads with long legs. Conversely, shorter-legged toads are much more likely to bump into other shorter-legged toads.

This "spatial sorting" can then enhance adaptations for other characteristics that make the long-legged toads even better dispersers. Say, a new mutation increases the ability of leg muscles to conserve energy, so the bearer of the mutation can hop for longer time. This imaginary adaptation benefits both short- and long-legged toads, but in a toad that is already a really good disperser—in other words, one with long legs—the benefits are much greater because it can disperse even farther. By not mating with short-legged toads, because there aren't any around where the long-legged toads are, the toads at the front of the invasion become even better dispersers.

That was some mental gymnastics. The point is our lineage started to split from the chimpanzee lineage because of the loss of a chromosome, which was at the time our ancestors were experimenting with coming down from the trees. Because the chimpanzee ancestors—those with 24 pairs of chromosomes—stayed in the trees, our ancestors, carrying 23 pairs of chromosomes, were much more likely to find a mate with the same number of chromosomes if they were spatially separated from the chimpanzee lineage. If, as I hypothesized earlier, it was also possible to discriminate between potential mates based on chromosome numbers, then not that many hybrids would have been produced, reducing selection against the chromosome-fusing mutation. Like choosing like was made easier because they tended to hang around each other.

So, how did the chromosome-fusing mutation get started?

It is an educated guess that it started with a male. Both chimpanzees and gorillas have a mating system in which a single male presides over the group, preventing, though not always successfully, other males from mating with females in the group. And that means a single male can sire a lot of offspring. Bonobos have a different, more egalitarian mating system, but for sake of argument I assume our early ancestors were more like common chimpanzees and gorillas today. (My assumption is not critical to the argument but does make it easier to envision how a chromosome-fusing mutation could spread in the beginning.) We know that the details of the mating system of a species do not make or break the separation into different species via chromosomal fusion, but harems do speed up the process.

Imagine our harem holder being the first individual with fused chromosomes. Assuming that the mutation that causes the chromosomal fusion is dominant, so that a single copy of the fusion mutation is sufficient to cause chromosomal fusion, many of the male's offspring will have 23 pairs of chromosomes. Our male's reign is long and at some stage some of his mates will be his daughters. In the meantime, quite a bit of conspiring and plotting has been going on and a few gung-ho youngsters join forces to topple the old male. One will be installed as the next dominant male. A relative of the old boy, the new harem holder will mate with his full sisters and half-sisters. The number of 23-pair individuals increases further.

This fictitious description could have come from Jane Goodall or Frans de Waal, both long-term observers of chimpanzee politics. There are other ways in which our chromosome-fusion mutation could have spread, but they all require that the initial group, or population, was small. When mating is restricted within a small group, a mutation can relatively quickly spread to all group members. And, yes, that means some level of incestuous liaisons was probably part of our evolutionary past.

Isn't incest supposed to be bad?

Darwin himself was concerned with the negative effects of inbreeding, having married the daughter of his mother's brother, his first cousin, Emma Wedgwood. Despite his concerns, they had ten children, of which three died before the age of eleven. Powerful European families, but also some ancient Egyptian pharaohs, had been marrying their close relatives for hundreds of years so as to ensure power stayed within the family. In some instances, that procedure went terribly wrong. Queen Victoria's family tree is probably the best-known example of such inbreeding gone wrong. She spread the mutation causing hemophilia (a serious blood disease) to ten of her direct descendants (out of sixteen), thereby spreading the rare disorder to the royal houses of Spain, Russia, and Germany.

Incest can be bad if the family is unlucky enough to have a really bad mutation, as Queen Victoria did, but sex actually creates a lot of variation. Remember that each parent gives to its offspring only half its chromosomes. And each chromosome "behaves" independently of every other chromosome, so for a single individual there are 2^{23} (2 to the 23rd power) possible ways to produce a gamete; an egg or sperm cell. That is 8,388,608 unique gametes per individual. But that is only the gametes from one parent. To produce an offspring, we need two gametes, so a human couple can produce 2^{46} (2 to the 46th power) unique offspring. That number is too large to write down here, but it explains why my brother is not like me at all. And as if that is not enough variation, sex has an additional mechanism to mix things up. When eggs and sperm are made by parents-to-be, bits that come from the chromosome of the father are mixed with bits that come from the mother. This mixing is called *recombination* and, as the word suggests, it leads to a new combination of alleles.

Over many generations, recombination results in chromosomes that carry genetic information from both parents, the four grandparents, eight great-grand parents, and so on. If you could color-code all your ancestors' chromosomes, you would see that they are a mosaic of the colors present in your direct family. Each colored bit represents a particular stretch of DNA, inherited from a particular ancestor, so if you start at the beginning of a chromosome and trace back from which parent each block came, you could have a stretch of DNA that came from your father's family, next to one from your mother's family, sitting next to a couple from your father's family, and so on. Sex truly shakes things up.

THE BEGINNING OF US

My admittedly speculative scenario—a small number of ancestors carrying a chromosome-fusing mutation mainly mating among themselves—helps explain why we currently find much lower levels of genetic variation among humans compared to chimpanzees, even though chimpanzees are restricted to a small part of the world and humans are found everywhere. Low levels of genetic variation can speed up evolutionary change. That might sound counterintuitive, as one often hears that low levels of genetic variation can push a struggling species over the brink to extinction, and that is most certainly a danger of low diversity. But in the case of our ancestors, the inability to breed successfully with what would become the ancestors of chimpanzees sped up adaptations to a permanent life on two legs, removed from the trees. If our mutual ancestors would have been able to mix freely, then any adaptation to bipedalism would constantly be diluted by mating with individuals better adapted to swinging in the trees.

All this is to say that our trajectory from tree-swinging ape-like ancestor to ruler of the world started with two chromosomes getting stuck together, leading to a small population in which weird stuff happened. That weird stuff included a slow-running molecular clock that held back our embryonic development, that gave us a head that sits proudly on its neck but with a baby face, and tiny changes to gene-regulatory networks that altered our skeleton so we became great walkers and runners.

Then, like a cherry on the cake, our brain became big.

four

MIND BLOWN

There's a lot you can do without a brain. Witness the scarecrow in *The Wizard of Oz*. Or, in nonfiction, one of my favorite organisms the slime mold. Neither plant, nor animal, nor fungus, a slime mold is a single-celled protist that lacks a brain or nervous system. Yet it exhibits a remarkable range of "intelligent" behaviors.

The slime mold has clear preferences when looking for food, favoring the cover of darkness over daylight. And like Hänsel and Gretel, a slime mold uses an external spatial memory to "decide" where to look next. Not by leaving bread crumbs, but by "remembering" where it has previously searched via the slimy substance it leaves behind as it moves through its environment. The critter can recognize danger, yet it also has learned to ignore danger cues if they stop correlating with bad outcomes. We have even been able to show that slime molds are capable of a kind of "irrationality" we usually think about only in connection with brainy organisms like us: relativity bias. How we rate an item is notoriously sensitive to the salient comparison class. That's why stores display stuff that's way more expensive than anything they expect you to buy. It is also why my husband and I could pull off a neat trick to sell our honey.

While working in the Bee Lab in Baton Rouge, Louisiana, many years ago, my husband had a colleague who was an aspiring artist with a love of bees. As part of an art project this artistic bee guy created the most stunning home for bees. That bee home still allows us to observe the bees while they go about their business undisturbed. Our observation hive has two glass walls and is made of solid timber, adorned with exquisite anatomical drawings of the innards of bees, of flowers, and with details of bees sucking nectar. This special bee home now shares our office, filling it with a gentle buzz and sweet odors.

Beautiful as it is, the observation hive is not without problems, as it provides the bees with only limited space. To avoid overcrowding we regularly need to take out a comb full of brood and swap it with an empty one so the colony never outgrows its tiny apartment. And for that task we need a support colony, one of normal size that we keep outside in an ordinary bee box and with which we can easily swap frames with the colony in our observation hive. (Beekeepers provide the bees with a wooden frame in which they can build their wax combs; we therefore normally use the word *frame* instead of *comb*.)

Before we moved to Australia's tropical north, we lived in a more benign climate perfect for honey production. And so our support colony quickly started to fill up with honey. Normally, honey is extracted from cells using a honey extractor, but having only one hive, we never bothered buying one. To solve our honey problem, we simply cut the honeycomb into pieces, put the pieces in containers, and gave the cut pieces to friends. But then we ran out of friends, and still the honey was flowing. The local shop provided a solution, offering to sell our honeycomb to the many people who came to this beautiful part of the world on weekends. We offered it to the shoppers in two sizes: The smallest comb pieces sold like hotcakes, as the shop's Canadian proprietor liked to say, but the medium pieces not so much. So, we needed to entice shoppers to buy the larger, more expensive combs so we did not have to spend so much time cutting and boxing small pieces. But how? By offering an even bigger and more expensive honeycomb option.

Now we had three sizes of honeycomb in the shop. And lo and behold, with the third option available, which was bigger and more expensive than the other two, shoppers now turned their preference away from the smallest honeycombs to the previously ignored medium option.

Why mention this? The shoppers behaved in an economically irrational manner, changing their preference when a third option became available that was worse, in terms of value, than the original two choices. Those irrational shoppers were in good company, as such behavior is well known from studies of a range of other organisms, such as blue jays, honeybees, starlings, and stickleback fish. But surely something that doesn't even have a brain could not likewise behave so irrationally?

Almost as a joke, I roped in an equally curious (or insane) colleague, and together we came up with an experimental test for irrational behavior in a brainless slime mold. The slime mold likes lots of ground oats mixed with agar (see it as the slime mold's version of a big honeycomb), but not if that oatmeal-agar mix is placed in the light (equivalent to a high price). By making different combinations of oatmeal-agar mix placed either in the light ("expensive") or in the dark ("cheap"), we could mimic a slime-mold version of our honeycomb experiment. To our astonishment, the slime-mold version gave us the same outcome: when offered a third choice that was worse than the two original choices, the slime mold changed its preference. It seems one doesn't need a brain to behave irrationally.

It was a fun experiment, but the real questions I wanted to ask my slime mold had to do with establishing the cognitive limits of brainlessness. That limit turned out to be *associative learning*. Apparently, if you don't have a brain, associative learning is impossible. This is because without the "hardware" to store the information, you can't learn anything and so intelligent behavior is limited to simple things like not foraging where you have previously foraged. I had reached the end of my adventures in slimy intelligence.

From no brain to a very big one. One of the biologically defining characteristics of our species, *Homo sapiens*, is having a seriously outsized brain. And the intricate wiring of the human brain has clearly given us a few legs up when confronted by various selection pressures. Of course, as we've seen, big brains are also one half of our species' original childcare problem. But as I'll soon show, those big brains were equally essential to solving that problem. We have come to the strange and surprisingly little-known genetic story of how our brains grew so large.

Since the start date of our species, roughly 300,000 years ago, the human skull has undergone some spectacular changes—changes far more significant than any developments in sister species of *Homo* in the millions and millions of years prior. The skull of *Homo heidelbergensis*, which was probably our immediate ancestor, is only marginally different in size and shape from the skull of *Homo erectus*, one of earliest members of the genus on record. Three serendipitous natural events built upon one another in quick succession to generate the explosion in neural capacity we exhibit today.

A chromosome "inversion" created what's called a *gene nursery* within the inversion. Then within the nursery, a long-defunct gene was repaired and regained its function. The final step occurred when the repaired gene became duplicated. The net result was yet another developmental delay—not neoteny this time, but a delay in cell differentiation within the brain. This delay provides more time for cell division, ultimately leading to more neurons. But let me elaborate on that thumbnail sketch.

AN HEIR AND A SPARE

The eighteenth-century English-born American political activist Thomas Paine is famous for his unflinching opposition to the British crown, where accession to the throne is acquired via birth, irrespective of the heir's mental or moral character. Paine referred to the hereditary monarchical system as a system of mental leveling that "indiscriminately admits every species of character to the same authority. Vice and virtue, ignorance and wisdom, in short, every quality, good or bad, is put on the same level. Kings succeed each other, not as rationals, but as animals." Unsurprisingly, Paine was not equally loved by everyone.

A hereditary monarchy ensures the continuation of the royal family's power if each king fathers both an heir and at least one spare. (Most monarchies traditionally were headed by a king and only sons would be in line for the throne.) If the heir to the throne meets some unfortunate fate before he has managed to produce his own heir and spare, the throne nonetheless remains safely within the family, passing as it must, to the spare. (That's the idea, anyway.) Being the spare relegates one to the role of second best to the older sibling, but this doesn't mean there aren't great advantages to having the position. The spare has the freedom to pursue alternative lifestyles (just ask Harry, the Duke of Sussex), while the heir is condemned to remain firmly within the royal straitjacket. In a similar vein, a "spare" gene created by a gene duplication has the freedom to evolve new functions because the "heir" gene will take care of the original. Like the monarch's spare, a duplicate gene is free to explore new functions that would be impossible for the original.

Imagine a gene—call it A—that codes for an important function in an organism through the production of an essential protein. Any change that scrambles the information encoded in the gene's DNA sequence, so it can no longer produce the protein it is supposed to produce, will quickly be selected against; any individual who has the misfortune to inherit a mutated version of the gene will most likely die or otherwise be disadvantaged. But if that is so, how can novelties or innovations occur? After all, novelties and innovations require changes to the genetic code. This is where the gene's spare can come in. When there are two copies of the same gene, the spare gene—A^1—is free to change, potentially leading to new functions. But that happens only if the spare stays in the population for long enough for those innovations to take place. And that, it seems, is far from certain.

In many instances, gene A^1 will disappear from the population through a process that geneticists call *genetic drift*. Because gene A, the heir, continues to perform its function as if there were no spare, gene A^1 initially has no real purpose; it is redundant. There is every chance that the spare gene A^1 will simply disappear from the population because natural selection has no "need" for it.

Take rabbits. Imagine a small population of rabbits—brown rabbits and white rabbits. Then, assume that to be a brown rabbit, the animal has to carry only one B allele, so that both BB and Bb rabbits are brown. (Remember, alleles are different versions of the same gene inherited from both parents.) In contrast, rabbits that carry bb alleles are white. When it's time to reproduce, it doesn't matter whether a rabbit is white or brown. In other words, there is no selection on either brown or white rabbits; coat color is selectively neutral. (In this example, that is, as one would never see a white rabbit in the wild unless it had escaped, leaving some poor child sadly bereft.) In the beginning, the two alleles, B and b, are at equal frequency in our imaginary rabbit population, but most rabbits are brown because just a single B allele is enough to be brown. We now let five rabbits reproduce successfully, and just by chance all of them happen to be brown (not surprising, of course, given that most rabbits were brown to begin with). Two of our lucky rabbits carry BB alleles, and the other three carry one of each allele, or Bb. In the next generation, the frequency of the B allele has increased to 0.7 (or 70 percent: two rabbits with two copies, plus three carrying one copy), and the frequency of the b allele has dropped to 0.3 (30 percent), simply by chance. Now, two rabbits from this generation successfully reproduce, and both happen to be BB rabbits, just because the B allele is more prevalent. And that means in the next generation there are no more rabbits that carry the b allele—the b allele has been lost from the population. The B allele, on the other hand, has gone to fixation, as every individual in the population now carries two copies of the B allele.

Natural selection has had nothing to do with the demise of the b allele, and by extension, the demise of white rabbits from this population. It's only pure chance. Through a similar chance event, a newly copied gene—the spare—often drifts out of the population because initially there is no reason it should be kept. But sometimes it sticks around anyway. That is when interesting things can happen. But before we go any further, we have to know something about the way chromosomes are structured.

We should start with flies—fruit flies to be precise. When I was a student in Amsterdam, our genetics lab was in one of those beautiful old buildings in the inner city. During our practicals with fruit flies, trying to figure out how many flies had red eyes and how many had white eyes, we temporarily knocked the flies out with ether. It being an old building not really fit for this purpose, the fumes would gradually build up until the instructors decided it was time for a break. They weren't particularly concerned we would work too hard, but they didn't want one of us to pass out while sorting our flies. We were sent outside while they opened the windows.

Our practicals weren't designed to look directly at the flies' chromosomes, but we felt we were at the cutting edge of human knowledge because the tests we were doing were made possible by decades of study by a large number of people. Alfred Henry Sturtevant was one of those people studying genetics in the early 1900s. Sturtevant was by all accounts a remarkable man with wide-ranging interests who saw "doing science" as a form of puzzle solving (which I guess in essence it is). But his real claim to fame lies in his many discoveries related to the structure of chromosomes. He also figured out that genes sometimes end up in weird places on chromosomes.

The humble fruit fly *Drosophila melanogaster* (some people will know it as the vinegar fly) has been an extremely useful organism for a range of reasons when it comes to modeling genetics. As anyone knows who leaves fruit in a bowl on the kitchen counter for too long in the summer, fruit flies are prodigious reproducers. Annoying as that is for ordinary folks, it is a godsend for geneticists. It means they can breed many generations of fruit flies in quick succession so as to look at the inheritance of characteristics (even in a school or college practical). The flies are also not too fussy about what they eat and how they are kept. Glass bottles with a cotton ball as a stopper will do. Feed them a mixture of cornmeal, sugar, and yeast, and they are happy as a lark. But there is more. The fruit fly has only four pairs of chromosomes, which makes figuring out which genes sit where on which chromosome much easier. It also has some tissues that carry many copies of chromosomes within a single cell—the cells are polyploid. Polyploid cells make it much easier to see things even with the primitive microscopes of the time (remember, we are talking the early twentieth century).

Sturtevant was the first person to map chromosomes—that is, to determine the sequence of genes along a chromosome. But some genes did not seem to behave as expected, thereby messing up his neat gene order. Remember from Chapter Three that during the production of sex cells—eggs and sperm—homologous chromosomes swap bits before they separate and are divided over the daughter cells. What Sturtevant found was that some genes moved completely to one of the chromosomes, leaving its sister-chromosome bereft of that particular gene. What was going on?

Another reason why the fruit fly is a geneticist's dream is that there are many mutant flies with specific phenotypes (remember the different eye colors from my school practical) that are easily maintained. The first catalog of fly mutations and their phenotypes was published in 1925. Now, *Drosophila* researchers can simply buy any available mutant strain they want. One mutation known in Sturtevant's time was the *Bar* mutation—a mutation that causes small eyes. This mutation is weird because it reverts to the normal gene way more frequently than expected. One normally expects this kind of spontaneous mutation to happen at a rate of one in a million (with a given gene per generation), but *Bar* fly females produce normal offspring—those with normal eyes—at a rate of one in a thousand.* How could that be?

Sturtevant was able to figure out that the changes to the flies' eyes associated with the *Bar* mutation weren't caused by a change in the gene but, rather, by changes in the chromosome structure around the gene. For some reason, crossing-over in the area around the *Bar* gene was unfair—"unequal crossing-over," in Sturtevant's words. Unequal crossing-over results in two copies of one gene on the same chromosome and loss of the gene on the other chromosome. Later geneticists found out that *Bar* flies have two copies of the normal version of the gene, and *ultra-Bar* flies three. It is now well known that unequal crossing-over is caused by a particular kind of selfish element, called *transposons*.

---

* To be precise, it is not the actual *Bar* mutation rate that is unusually high but, rather, its back-mutation rate.

Transposons are bits of DNA sequence that can jump around the chromosomes, often taking other bits of DNA with them—like the *Bar* mutation in the fruit flies. Because they jump around, transposons are also known as "jumping genes" and were first described in corn by Barbara McClintock in the early 1950s. McClintock had figured out how to stain each of the corn's chromosomes differently so she could follow how the chromosomes lined up during cell division. To her surprise, she soon found out that bits of one chromosome often ended up on another chromosome. Instead of being fixed onto the chromosome, the genes could jump! McClintock's discovery was not immediately celebrated, but that didn't stop her from continuing her work. It took thirty-three years for the scientific community to recognize the importance of her work. In 1983, McClintock received the Nobel Prize in Physiology and Medicine for her discovery of jumping genes.

Transposable elements and the unequal crossing-over they cause are an important force driving evolutionary change. Unequal crossing-over can generate copies of genes that can then take on a new function—the spare to the heir—or even create whole gene families. The *Hox* gene family discussed in Chapter Three is an example of such a gene family. As it turns out, the human genome is rife with transposable elements, causing "mistakes" left, right, and center. Luckily, when it comes to evolution, mistakes can be a force for good.

A big mistake happened on our chromosome 1, probably just after our split from the ancestor that our lineage shared with chimpanzees. A piece of chromosome 1 broke off and had to be reaffixed. But the repair job got botched, and the broken bit was glued back on upside down, forming what geneticists call an *inversion*. An inversion doesn't lead to any changes in the genetic information, just a reversal of gene order. (Recall the mystery of the 1 percent in gene sequences between chimps and humans.) Sturtevant was the first to describe an inversion in his fruit flies, way back in 1921, when he was still an undergraduate student. We now know that many events like Sturtevant's *Bar* mutation happen within or around inversions. In effect, inversions create gene nurseries, or areas on chromosomes where gene duplications are frequent and often lead to the spare taking on a totally new function through "gene conversion." The inversion on our chromosome 1 contains a disproportionate number of genes that distinguish the human lineage. One gene in particular deserves special attention. That gene is Neurogenic locus notch homolog 2, or *NOTCH2*, a member of the *Notch* gene family responsible for cell signaling.

MIND EXPANDING

Like all living things, cells communicate with other biological units in their environment by receiving, processing, and transmitting coded messages. *Notch* gets its name from the effect a particular mutation in the gene has in the fruit fly: It causes a notched pattern on the flies' wings. (The fact that fruit flies and humans share the same gene families can be explained only through our common ancestry; remember that when you are next chatting with an evolution denier.) Members of the *Notch* gene family have many fingers in the pie when it comes to development, but there is one role to focus on. And that has to do with *neurogenesis*, the production of neurons from neural stem cells via radial glia cells.

Stem cells are undifferentiated cells, often called *totipotent cells*, that have not yet started their trajectory to a particular cell type. See them as a blank slate, waiting to be guided to their final form by their environment. The *Notch* gene family orchestrates the *Notch* signaling pathway, which in turn influences the differentiation of stem cells. The regulation of cell fate—in other words, when a stem cell starts to specialize and what it will specialize in—is highly complex. Different genes produce products that act in synergy, complement, or even counteract each other. The first "decision" a stem cell needs to make is whether it will continue to proliferate, thus producing more stem cells, or will start to specialize. In the brain of vertebrates, stem cells are called *radial glia cells*. Radial glia cells can decide to copy themselves and produce more radial glia cells or start to differentiate into neurons. That decision—to divide or to differentiate—is influenced by Hes factors, proteins produced by *Hairy and enhancer of split*, or *Hes*, genes. We have met one such gene, *Hes7*, in the previous chapter; Hes7 proteins play an essential role in the central pacemaker that orchestrates embryonic development. *Hes* genes code for proteins that suppress transcription either of themselves or of other genes.

In the mammalian brain, Hes factors tell radial glia cells it is time to start differentiating into neurons through a series of other cell types. Differentiation is a one-way street; a neuron cannot turn back into a radial glia cell. The number of radial glia cells in the brain determines how many neurons can eventually be produced. As a radial glia cell transforms into a neuron, it migrates from the inner core of the developing brain out toward the periphery until it finally reaches the cortex. Pictured en masse, this cell migration looks a bit like a balloon inflating in slow motion.

When there are no Hes factors, radial glia cells continue to copy themselves, producing more and more radial glia cells. You can see where this is going. A delay in radial glia cells being told to differentiate results in more neurons in the cortex, because there will be more radial glia cells that will ultimately become neurons. All these extra radial glia cells need to go somewhere, and as they travel outward to the brain's periphery, the brain expands.

I haven't forgotten about *NOTCH2*. The role of *NOTCH2* is to regulate how much Hes factor is produced; when *NOTCH2* is expressed, its products suppress the expression of the *Hes* gene responsible for producing Hes factors and the brain grows. Thus, *NOTCH2* is indirectly responsible for the growth of the brain. How do we know? Through an ingenious set of experiments using cell lines. Cell lines are a population of cells that come from a particular organism and that can be kept for an extended period of time. Probably the most famous, and oldest, cell line is the HeLa line, a human cell line that came from the cervical cancer tumor of Henrietta Lacks.*

Mice do not have the human version of *NOTCH2*, but it is possible to trick mouse cells into expressing the gene and then look for biological markers of neuronal differentiation. Similarly, one can prevent human cells from expressing *NOTCH2*, using CRISPR-Cas9, and look at what happens to the same

---

* Henrietta Lacks's tumor cells were cultivated without her consent in 1951. She died later that year. Her family learned only many years later that Lack's cells had some specific characteristics that made it possible to establish the now famous cell line. The medical profession only recently admitted the injustice done to Lacks and her family; see https://www.hopkinsmedicine.org/henrietta-lacks.

markers. CRISPR-Cas9 is a molecular scissor technique for which Emmanuelle Charpentier and Jennifer Doudna received the 2020 Nobel Prize in Chemistry. Mice cells with the human *NOTCH2* version showed evidence of delayed neuron differentiation; human cells that did not express the gene did not delay neuron differentiation.

Of course, humans are not the only organisms that have neurons, but we do have a prodigious number of them. And that has everything to do with the spare of *NOTCH2* that popped up about 8 million years ago, before the human, gorilla, and chimpanzee split. The spare became *NOTCH2NL*. It's a spare that did nothing because in modern gorillas and chimpanzees the spare has no function at all—it is a pseudogene. But that all changed in our lineage 4 to 3 million years ago, thanks to the break in chromosome 1 and the resulting inversion. *NOTCH2NL* was revived! Not only was it revived, it also made two copies of itself so we now have three functional versions of *NOTCH2NL*. In the last few hundred thousand years, this triplet underwent more changes, and now acts on *NOTCH2* so that *NOTCH2* makes even less of the stuff needed to set the radial glia cells on the path toward differentiation.

Thanks to the revival and then the triplication of *NOTCH-2NL* owing to accidents in our genome, our species' brain expanded because of the delay in maturation of the precursors of our neurons. But there are downsides. People with only two copies of *NOTCH2NL* have a much smaller brain (microcephaly), while those with one copy too many have a much larger brain (macrocephaly). Both situations are known to be associated with neurological abnormalities. It seems that as a species at least, we can't have our cake and eat it, too.

But there is more to the expanding-brain story—and it gets a little creepy. Other cells also add to evolution's human brain–expansion project. Neuroepithelial cells form the innermost layer of the mammalian brain from which radial glia cells develop. (*Epithelium* is a generic term for a kind of tissue that forms a sheet of cells, protecting tissues and lining cavities in all animals. A special type of epithelial cell makes your skin, for example.) Like all epithelial cells, neuroepithelial cells stick together through a particular protein that basically acts as a glue. Before neuroepithelial cells can turn into radial glia cells, that glue needs to be dissolved, freeing the cells to start their transformation. A delay in dissolving the glue would then result in more neuroepithelial cells, because each time a cell divides it gets caught in the epithelial layer when the glue is still around. And more neuroepithelial cells leads to more radial glia cells, which expands the brain. Great story, but how do we know it is true?

By growing brains in a petri dish. Yes, you read that correctly. In a slightly creepy experiment, researchers were able to grow mini brains in the lab from human, chimpanzee, and gorilla cell lines so they could mimic the early development of brains. It was pretty much as if they were able to look at very young embryos. After five weeks, the human miniature brains were around twice as large as those of chimps and gorillas at the same age. And that had everything to do with the expression of a gene that affects the amount of "glue" holding the cells together: zinc finger E-box binding homeobox 2, or *ZEB2*. A delay in the expression of *ZEB2* keeps neuroepithelial cells together for longer, leading to more radial glia cells and a larger brain—or, mini brain in the case of this experiment. When the researchers switched on *ZEB2* earlier in the human mini brains, they looked more like gorilla mini brains. When they delayed the expression of *ZEB2* in the gorilla mini brains, they became more human-like.

Both *NOTCH2NL* and *ZEB2* illustrate again how a relatively simple change—gene duplication and a delay in gene expression—can have a huge effect. We are getting closer still to solving the mystery of the 1 percent.

THE IMPORTANCE OF STAYING COOLHEADED

Many years before I moved to Australia I came for a visit, combining a scientific conference with a holiday. During the holiday part of the trip my family and I decided to walk a famous track on the uninhabited island of Hinchinbrook. In a funny twist of fate, I can now see Hinchinbrook from my local beach, but at the time I was one of the many European travelers to Australia's tropical north. And as with most European visitors, I had no clue about the tropics, so I headed off on the four-day walk without a hat. Parts of the track are in the blazing sun, and soon I started to feel weird. I felt tired and a little dizzy, my heartbeat went up a notch or two, and it seemed as if I was going to be sick. Luckily I quickly realized what was going on: my head had gotten too hot from the sun and lack of protection. Resting a little in the shade while drinking loads of crispy cold water from the stream nearby soon made me feel better. The walk I finished with a spare pair of shorts on my head.

It could have been so much worse had I not recognized the early signs of heat exhaustion, which when ignored can culminate in potentially deadly heatstroke. The brain is extremely sensitive to high temperatures, and at a core body temperature over 104°F (40°C), it is in danger of being permanently damaged.

(Fevers caused by infections hardly ever cause life-threatening temperatures.) But there is another reason why my holiday trip did not end badly. I carry multiple copies of the gene *Aquaporin 7*, or *AQP7* for short. Not just me, but all of us. And thanks to that, we sweat. A lot.

A rise in body temperature activates the sweat glands, which then bring water mixed with some salt to the surface of the skin. As the sweat evaporates, we cool off because the conversion of a liquid to a gas costs energy in the form of heat. Another way to get rid of excess heat is by panting, as anyone with a dog knows. Some other animals also sweat (think horses), but they mainly rely on panting to avoid overheating.

While we pant during exercise, the panting does not cool us off but, rather, increases the intake of oxygen and helps in getting rid of carbon dioxide. We are so effective at dumping excess heat by producing sweat because we have an enormous number of sweat glands—many more than any other animal. One part of the body has a disproportionate number of them: the head. Effective thermoregulation through sweating prevents the brain from overheating so we can be out and about even under strong radiant heat (up to a limit, of course, as my ordeal illustrated).

The paleoanthropologist Daniel Lieberman thinks that the ability to run long distances while keeping the head cool allowed our relative *Homo erectus* to become a hunter and thereby increase its brain size. Because other animals quickly overheat when running, *H. erectus* could hunt even much larger animals by continuously chasing them until they died of heat exhaustion. If Lieberman is right, and I think he makes a compelling case, then *H. erectus* most likely already had multiple copies of *AQP7*. The *AQP7* regulates the movement of water and glycerol across *aquaporins*, or channels through the cell membrane, and so affects the amount of sweat produced and mobilizes the fat reserves needed to fuel the running. One gene, two functions. More copies, better functionality. Carrying multiple copies of *AQP7* (we have five) then allowed our species to cope with an ever-expanding brain when *NOTCH2NL* got repaired and duplicated.

AQP7 does not sit on our chromosome 1 but, rather, on chromosome 9, another chromosome that is rife with duplicated genes. Apparently we humans have a chromosome architecture that encourages weird stuff like conversions and duplications. That is not to say that gene duplications are unique to humans, but some of our chromosomes are clearly extremely fruitful gene nurseries. Those nurseries have delivered small genetic changes with huge effects.

MOLDING A HEAD

If you were to look at the 300,000-year-old *Homo sapiens* fossil found in Jebel Irhoud, Morocco, you would see a face that looks rather familiar but a head that does not. The Jebel Irhoud fossil no longer had a snout, but it has a face much flatter than that of our Neanderthal cousins. What the skull lacks is the bulbous shape so typical of us modern humans. That change took another 200,000 years or so, until we start to find fossils that are indistinguishable from modern-day humans. You may wonder why such diverse creatures are lumped into the one species, but all share that one typical *H. sapiens* feature that has been mentioned so often: a very large brain. And that brain changed over a few hundred thousand years, causing the whole skull to change, too.

In a way, the skull seems a bit of a contradiction in terms. It is not one thing. It is a complex structure, comprising many different bits of bone mixed in with cartilage, tendon, ligaments, enamel, dentine, and cementum, all held together by myriad muscles. For all this complexity, the skull is extremely versatile. Versatile in an evolutionary sense, that is. Maybe the skull is so versatile because it is so complex, as its mosaic structure allows natural selection to play with the final product, molding it to the precise requirements at the time. And there are many requirements. The skull is much more than a protective case for the brain. Almost everything we do requires the head, in one way or the other. Just think about it (pun intended). We use our head for chewing, swallowing, tasting, seeing, hearing, breathing. But the head is also important for less intuitive activities, such as thermoregulation (by moisturizing and warming the air we breathe) or moving and balancing (by keeping runners from stumbling and falling mile after mile).

Our ancestors' diet most likely had an important influence on the face by way of the effect of bite force on the shape and position of the jaw and the muscles needed for chewing. Chimpanzees spend about 50 percent of their time chewing because they need to eat a lot to satisfy their nutritional requirements, mainly depending as they do on fruits. Because of the inclusion of meat in our ancestors' diet, started by *H. erectus*, it became easier to get sufficient calories while eating less, and thus chewing less, even prior to the invention of cooking. (Cooking food was probably invented only after our species appeared, although we don't really know for sure.) Meat can be pounded to make it softer, and thus easier to chew. The shift in diet therefore probably helps explain why later *Homo* species look much less chimpanzee-like when you look at the face. However, there is more to understanding why our head became so round.

Think back to the effect of neoteny on the shape of the skull. When explaining neoteny and its effect on our head, I made a direct comparison between the shape of our (modern) skull and that of chimpanzees—the difference came from the slowing down of a molecular clock, which in turn affected the skull's growth rate. But, wait—the skull is a complex, multifaceted component; how can all its parts slow at exactly the same growth rate? Compared to the growth of the skull of a chimpanzee, the rate of growth of different parts of our skull is highly variable, so that different components grow at different rates. While skulls are constructed of many different bits, we can divide the skull into three main components: the face, the cranial vault (the space that encases and protects the brain), and the cranial base. Each is heavily involved in the other two components, so that when one changes, the other two must also change to accommodate the change in the first component. So, when selection drives one skull component in a particular direction, then the rest of the skull will follow in a "buy one, get a second and even third one free" kind of way. And that, in turn, suggests the differences we find between our own skull and those of chimpanzees are due to more than a simple slowing down of a developmental clock. It appears that our skull has been under selection pressure that is far different from that of the skull of our close relatives.

It is easy to claim the three different components of the skull act as one, but how do we know? We know because there's a room filled with skulls. That room is the ossuary of Hallstatt, Austria, also known as the Hallstatt Beinhaus (*Beinhaus* is German for "house of bones").

Hallstatt is an idyllic village positioned on the shore of Lake Hallstatt, surrounded by the Salzkammergut Mountains. Idyllic as Hallstatt might be, in the 1700s the local church started a less than idyllic tradition. They started to dig up corpses of people who had died in the the preceding ten to fifteen years. This somewhat macabre practice had a pragmatic reason: The cemetery was getting too full. And so family members of the deceased were asked to stack up their loved ones' bones inside the house now known as the Hallstatt Beinhaus. The bones of relatives would be placed next to kin, thereby starting a genealogical record. From 1720 onward, the skulls were adorned with symbolic decorations, as well as the dates of birth and death. The practice of digging up the remains and keeping the bones in the Beinhaus continued until the 1960s, providing interested scientists with a wealth of data (assisted by the church's own records) to figure out how characteristics of the skull are inherited.

Those scientists included Neus Martínez-Abadías and her colleagues. Using the skulls of 390 deceased adults born between 1707 and 1885, Martínez-Abadías's research team characterized each skull shape with twenty-nine anatomical landmarks on the left side of the skull. They used these landmarks to measure the variations in shape and size of the different parts of the skull. It was then possible to determine how a change in one of the three main components—the face, the cranial vault, and the cranial base—would affect the other two. If the different components are independent, then one would not expect to find that changes in, say, the face have a consistent effect on the cranial vault and cranial base. If, on the other hand, a particular change in the face always leads to a specific change in the cranial vault and base, then we can assume the three major components of the skull are integrated.

The team also used a clever way to gain insights into the effects of selection on each component of the skull independently. For this, they made use of the familial relationships among the people whose skulls they used. (Of course, records based on whom people think they are related to are not as reliable as DNA would be, but in science, like many other things, one often needs to make do with what one has.) From the findings, the researchers could infer that during the evolution of our species the three components of our skull changed in a particular direction. The cranial base flexion increased (this means the angle between the line you can draw from your eyebrow across the top of your ear and then down to the base of your skull becomes smaller; remember the figure from Chapter Three), the face becomes flatter, and the cranial vault expands and becomes more rounded.

To answer if a larger cranial vault would also lead to an increase in cranial base inflexion (reducing the angle between the two fictitious lines on the side of the head) and a flatter face, thereby explaining the changes to our skull by an increase in brain size alone, the only other thing the team needed was a method to quantify such potential response to selection. Fortunately, such a method was developed in agriculture long ago to answer questions such as "Does picking cows that produce a lot of milk result in offspring that produce even more milk?" That method is known as the *breeder's equation*: a method that allows one to determine the response to selection of a particular phenotypic trait and its heritability. Or, more precisely, the multivariate breeder's equation, as here we are looking at more than one phenotypic trait and the ways selection potentially acts on them.

What were the results? I've given away the punchline: changing one component changes the others, too. The data suggested strongly that the different components of the skull were not independent. It seems reasonable to assume that our bulbous head has everything to do with the need to accommodate an expanding brain.

AN EVOLUTIONARY SPATIAL PACKING PROBLEM

For quite a few years, I was interested in finding computational solutions to packing problems. Not because I had a particular issue with packing bags before going on trips (I still do), but because I wanted to see if I could find inspiration from social insects to reach better solutions. (I blame sharing an office with a computer science PhD candidate for my digression.) At the time, in the early 2000s, biologically inspired optimization algorithms were all the rage. Optimization algorithms find a "good enough" solution to problems that are almost impossible to solve analytically. Probably the most famous example of a so-called hard optimization is the traveling salesman's problem. Before the invention of television advertisements, or any other means of advertising, salesmen (as I doubt there were any women doing the job) traveled from door to door with a sample of the goods they were representing—vacuum cleaners, encyclopedias, what have you. They often traveled to different cities, and in doing so wanted to find the shortest routes between their cities while avoiding wasting time visiting the same city twice. That shortest-path problem became the traveling salesman problem, a benchmark for computer scientists testing their optimization algorithms. But then someone realized there are creatures that have been solving such shortest-path problems for millions of years: trail-laying ants.

A new field of bio-inspired optimization algorithms came into existence, taking their cue from the ways ants solve their traveling problems, and I found myself spending time with someone doing a PhD in that field. We became interested in packing problems—in particular, the knapsack problem—which was another benchmark question often posed in computer science: how to pack as many individual items as possible into a confined space. For a while, we tried to see if we could use honeybees as our inspiration. Instead of packing as many identical items, the knapsack problem asks you to pack items that differ in value, with the ultimate aim of maximizing the total value of your knapsack's contents. We spent some wonderful times (in the office and in the pub) brainstorming to see if the way bees forage—collecting nectar and pollen from a range of plant species that differ in their quality—could give us inspiration for how to pack a knapsack. Alas, we failed. But we had a good time trying. And I did learn a lot about the importance of packing effectively—in theory, at least.

Back to the human skull. It is all good and well to argue that we can explain our typical *H. sapiens* head with the need to pack that large brain well, but it would be better if we could somehow test the idea experimentally. It turns out we can, using mouse strains with different mutations that affect the growth of their skulls. Here's the reasoning behind the experiment. Because the brain grows on top of the cranial base, a more flexed cranial base allows more volume without the need for any other changes (the great advantage of the skull's interconnectedness). The study used three mutant mice strains to test the spatial packing hypothesis. Each mutant changed part of the head in such a way that, if spatial packing is *the* explanation for the shape of skulls, then each mutant would affect the mice's head in particular ways. And indeed it did. The mutant mice showed that the angle of the cranial base (the cranial flexion) helps the cranial vault accommodate the volume of the brain and at the same time it affects the face. The smaller the cranial flexion, the more brain can be fitted in and the flatter the face is. One change, multiple consequences.

To me, it makes a lot of sense that the skull is so malleable, prone to easily accommodate major changes, precisely because the head is such an important part of the body. (Also, just think about how the head of an individual changes from baby to adulthood.) Better provide for the change or fail otherwise. It's natural selection as the master tinkerer again. Such tinkering is possible because of developmental shifts (an expanding brain needs space) and biomechanical forces (change in diet) placing direct selection pressure on the skull. At the same time, it seems highly likely that the development of the brain and the skull are under the influence of the same molecular signaling pathways. We don't know exactly how that works in humans, but research on birds gives some intriguing insights.

If you have ever seen a cassowary (a regular visitor to my garden), you see immediately that modern birds are living dinosaurs. A cassowary has many dinosaur features: massive legs (they can no longer fly), a skin-covered casque on its head, and footprints that are indistinguishable from dinosaur footprints. And, frankly, they don't seem to be too intelligent.

Modern birds are an immensely diverse group. As Darwin discovered, many birds evolved diverse and highly specialized beak shapes in response to their environmental conditions. In the case of the Galápagos finches, a group that is also termed "Darwin's finches," the birds' beaks adapted to the prevailing seeds found on their specific islands after one species left mainland Ecuador and its subsequent ancestors spread across the Galapágos Islands. A bird's beak is in essence a modified reptilian snout, but to understand how we get from a dinosaur snout to a bird beak, we can look at two more easily obtainable model organisms: the chicken and the mouse.

If you are a clever molecular biologist who also knows how to study embryos, you can figure out which signaling systems are activated in which part of the embryo at what time during its development. Doing so reveals that the markedly different facial features of chickens and mice are under the influence of the exact same signaling pathways (of which our old friend Sonic Hedgehog is one), but are expressed at different times and in slightly different parts of the embryo. Of course, I am grossly simplifying here, but the point is that, like the rest of our skeleton, the skull is easily adapted, thanks in part to evolutionary conserved gene networks and signaling pathways. No novel genes are required. And that implies the seemingly massive differences among the skulls of chimpanzees (our relatives), our cousins the Neanderthals (archaic humans), and us aren't that massive when we realize that changes in the expression of just a few developmental genes can cause such huge changes in skull structure.

One last thing on brains before we move on to where all these genetic flukes and molecular fine-tuning led. Neanderthals had almost the same-size brains as modern humans, yet their head shape did not go through the same modifications as ours did. Their cranial base remained less flexed, and thus did not create space to optimally pack in a large brain. Shape-wise, their heads were similar to those of archaic humans like the 300,000-year-old Jebel Irhoud I mentioned earlier. Because of our more flexed cranial base, though, our face is shorter than that of both Neanderthals and archaic humans. And that short face has had one monumental anatomical consequence.

PART TWO

... And Then We Started Talking

five
LOUD MOMS

Ever wondered how a nursing infant can breathe and drink at the same time? They never seem to get milk down the wrong spout. Unlike babies, adults are always at risk of choking on food or drink—so much so that choking on food is apparently the fourth leading cause of accidental death in the United States. We are prone to choking because we have a design fault that takes about four months to kick in. As our brain grows larger, our throat elongates, causing the larynx to descend and detach from the epiglottis—a flap of cartilage that seals off the windpipe while we eat and drink. This dangerous defect was an upshot of architectural changes to the skull required to accommodate an expanding brain. But the descent of the larynx had an unexpected consequence that helped it overcome steep selection pressures to earn its keep. Our remodeled throat allows us to make sounds like no other animal. It lets us speak. That design fault is probably the best thing that could have happened in our evolutionary history.

But what *is* speech?

In essence, speech is nothing more than puffs of pressurized air from the lungs that move up the windpipe (the structure that connects the larynx to the lungs) and strike the underside of the vocal cords, producing vibrations. The vocal cords, in turn, are large folds in the mucous membrane that lines the larynx. The grunts that emerge from the larynx enter the resonating chamber, the oral cavity. The final sound is then molded by the tongue, teeth, hard and soft palates, and lips as it passes through the mouth.

Making sounds is, of course, not restricted to humans; the world would be a boring place if that were the case. On rainy nights I can hardly hear myself think because of the chorus of many species of frogs in the creek that runs through our property. Frogs have skin flaps in their throats that serve as resonating chambers. To croak, a frog inflates its lungs, closes its mouth and nose, and shunts air through its larynx into the vocal sac. That's a great way to make a racket, but articulate speech calls for more control. This is where the tongue comes in, as anyone who has had an anesthetic at the dentist has found out. (Don't try to participate in a discussion just after a dental anesthetic, as I once foolishly attempted.) Without fine-tuned command of the tongue, we wouldn't be able to create "quantal" vowel sounds like "i" and "u," which linguists teach us are key to precision speech. As infants, our tongues were just for sucking, so they sat almost entirely in our mouths—as they do throughout the life spans of other apes and monkeys. But as we grew, our larynx dropped and pulled the tongue downward deep into our throat, not only increasing the size of the vocal cavity but also giving us more flexibility to mold sound. Our tongue is so flexible it can be moved up, down, forward, and back; bunched up or extended; widened or curled. And that flexibility matters, because when the tongue changes shape, the whole vocal tract changes too, altering the sound. If you ever wondered why severing the tongue of one's enemy was such a widespread practice, now you have the answer. Without it, or without complete control over its movement, a person cannot speak. No speech, no heretics or betrayals.

One of the birds I hear almost daily around our house is the noisy pitta, a small colorful creature that scuttles around on the forest floor. I usually have a hard time remembering bird calls, but ever since I saw the pitta's call described in a field guide as sounding like "walk to work," I can't forget that bird's call. And I even imagine that I can hear the words! Alex, an African Grey parrot, was probably the world's most famous parrot, learning to ape over a hundred English words, a feat that garnered him an obituary in *The New York Times*. Reportedly, the last words Alex uttered to his trainer were, "You be good, see you tomorrow. I love you." As with the noisy pitta, it takes a lot of wishful thinking to find meaning in these soundwaves. But complex phonology is an important precondition for complex semantics. We can't hope to express that it may rain or sleet next Tuesday unless we can recombine enough atomic sounds (or shapes) to carry that meaning.

Parrots are expert copycats, thanks to a proportionally larger brain compared to pittas and most other birds. Add the right morphology to the mix, and voilà, a recipe for speech. (Lacking lips, though, parrots still struggle with consonants.) Humans stumbled onto the right morphology thanks to the changes in our head and face necessitated by our ballooning brains. But why did our ancestors use their newfound cognitive and vocal powers to develop something as strange and unprecedented as syntax?

Generative grammar and compositional semantics are singularities in the tree of life. If evolution is so frugal, why did it bother with language? What did humans need to do that birds didn't?

Care for premature offspring. Exceptionally premature offspring. Our original childcare problem arose at precisely the evolutionary moment—150,000 to 100,000 years ago—when we finally had the anatomy and neurology to talk our way out of it. Because the problem and solution shared a common cause: our outsized brain.

We are not unique in our need to care for offspring. In fact, stepping back a little and looking at nonhuman animals is quite illuminating. So, let's stick with birds for just a bit longer.

A BRAINY PROBLEM

Birds can tell us something about the problems linked with having a big brain. They can also help us answer an evolutionary puzzle: If having a big brain leads to a range of cognitive benefits, and many studies have shown this to be the case, then why doesn't everything have a large brain? Especially given that studies have also shown that having a large brain is linked to living longer. The answer may be that not everyone can afford growing one.

Remember that a brain is very expensive; you need to eat a lot of the right food to be able to grow the brain. But while you are growing your brain, you don't yet have the cognitive benefits, just the costs. A typical catch-22 situation. And that means many species cannot afford to grow a large brain even if that would give them a huge leg up as adults. This idea, that brain size is limited because of its costs, can be tested in birds.

Birds roughly come in two kinds: those in which the young

need to look after themselves as soon as they hatch (precocial species such as ducks and chickens); and those in which the young need to be provisioned, often for extended periods of time (altricial species like pigeons). What has puzzled many for a long time is that all species with a large brain, like the parrots, Caledonian crows, and many others, are altricial. That is interesting; could it have something to do with having others care for you when you are young, helping you grow your brain? That is exactly the idea a bunch of evolutionary biologists decided to test.

Not only do birds come in two kinds, but it is also fairly easy to measure how much is invested in each offspring. One can measure egg size as a proxy for the amount of energy the mother puts into the developing embryo. Because birds are fed directly, one can monitor how many feeds each juvenile bird receives. Then, there are many species that breed cooperatively, where multiple adults look after young birds. Such alloparents—adults that are not the parents of the offspring they provide for—can boost the amount of resources put into raising the offspring.

Looking at 1,176 species of birds for which reliable data were available, Michael Griesser and his colleagues found there is one strong predictor of brain size: the amount of energy invested in each individual young. Large brains can evolve in species whose parents are able to gather enough food to sustain themselves and to feed their offspring's hungry brain. Having a large brain, then, comes with increased cognitive abilities, making the parents even better providers. A beautiful case of a positive feedback loop.

I love the simplicity of the causal relationship between the amount of energy available to young and the brain size attainable as an adult. It also explains some other patterns researchers have been focusing on, beyond avian species. Probably the most well known is the social brain hypothesis—the idea that a large brain is required to process the information needed to maintain social relationships. The social brain hypothesis provides an explanation for the positive association between brain size and social complexity across primate species. To live in a social group requires the cognitive skills to navigate complex social relationships and so, it is thought, natural selection drives an increase in brain size in social species.

The social brain hypothesis makes intuitive sense. An example I used for many years in my lectures on animal behavior featured two young chimpanzee males who had struck a deal. Remember that common chimpanzees (not bonobos) live in a group reigned over by a single male. That male does not want his underlings to mate with any of the females in the group, but the young males have a different opinion. Females, too, are often quite happy to mate with males other than the dominant one. The two youngsters in my example formed an alliance designed to get access to a willing female without the dominant male knowing. But how to keep the much stronger dominant male away from the mischievous pair? For that, the two young males took turns. One day one of them would distract the older male by teasing him until the dominant male became so enraged he charged after the buffoon. That was all part of the plan, as the teaser could now lead the male away from where the secretive pair were hanging out. The next day the roles were reversed, so both young males had a chance to mate with the female. That is, the two young males engaged in reciprocal altruism: I'll scratch your back if you scratch mine. Which is quite remarkable. Not only did they need to have the cognitive skills to come up with the scheme in the first place, but they also had to strike a deal with their partner-in-crime to reciprocate. That, in turn, required trust. Dreaming up the scheme and successfully pulling it off require a big brain. A social brain?

Is the social brain hypothesis the reason *why* brains are larger in social species? I think the bird study tells us no. Big-brained species are social because they can be. Brain size and sociality reinforce each other. A large brain provides the cognitive abilities required to build and maintain social relationships, but only because the social relationships ensure there are others to care for the young. It then follows that in all big-brained species, youngsters grow up in an environment replete with social interactions. It's an environment highly conducive to further sharpening the cognitive abilities. A big brain needs others, and the presence of others further shapes the big brain.

SELF-DOMESTICATION

I once had a dog who became pregnant without our knowing until it was way too late. And so, one day, the whole family—and that included the four puppies—went on a weeklong holiday to one of the islands north of the Netherlands. We were lucky that the accommodation we booked allowed pets—no one was going to look after the whole canine family if we left them at home—but traveling with a box of four puppies made our "relaxing" holiday rather challenging. Particularly for mommy dog. Normally she was quite a relaxed little dog, but all that changed the moment her pups were out of sight. As soon as we arrived at our accommodation, and we could reunite the young family, she gathered them up, curled around them protectively, and gave us the dirtiest look possible. (I made that last bit up, but I could imagine what was going on in her head.) No one was allowed to get near her puppies. She begrudgingly allowed us access if we needed to clean their box, but we were always alert to the possibility she would lash out.

The contrast between my dog's attitude toward her puppies and that of a human mother cannot be more extreme. My first introduction to the baby sister of a boyfriend was when her mom pushed her into my arms. "Here," she said in jest, "you look after her for a while." The girl was tiny, only a few weeks old. I had never met the mother before.

Both moms, canine and human, behaved in a species-stereotypic way. But it is the human mom who is the aberration compared to our great-ape relatives. No orangutan, gorilla, or chimpanzee mother will ever hand over her dependent baby to someone else. And no orangutan, gorilla, or chimpanzee baby would be relaxed about being handled by a stranger, as my boyfriend's baby sister was. The difference between us and the other apes is not that our babies are less dependent on mom; it's just that human moms simply need all the help they can get. And for that help, they must trust others to care for their dependent charges. Even strangers.

Sarah Blaffer Hrdy starts her book *Mother and Others* with a thought experiment: What would happen if you fill an airplane with chimpanzees instead of humans? Mayhem is the short answer. There is no way you can put a bunch of chimpanzees that don't know each other together in a confined space for any length of time, let alone hours. It would be a bloodbath. Yet, many of us humans do just that on a regular basis. We get onto a plane, or in a train or bus, with complete strangers, and nothing bad happens. We get to our destination with everyone intact. How come?

We have domesticated ourselves so we became less aggressive to our fellow human beings and more cooperative. The question is why we did this. The most likely driving force behind human self-domestication, as the hypothesis is known, was the need to share resources. When you can't do it on your own, your only chance to survive is to team up with others. And for that you need to tolerate and trust the others. Our other cousins, the bonobo chimpanzee, are a totally different beast from common chimpanzees. Bonobos, too, are less aggressive to others, even complete strangers. Whereas a common chimpanzee would try to kill a member of another group, a bonobo is more likely to have sex with the stranger. Bonobos split from common chimpanzees around 2 million years ago, probably when a population crossed the Congo River during a dry spell. (To date, bonobos are restricted to south of the Congo River, in one of the world's most unstable and dangerous countries, the Democratic Republic of Congo.)

The most prominent explanations for why we domesticated ourselves have to do with the problems of getting enough food. We know the environment changed about the same time our ancestors started to walk on two legs, and food may have become more difficult to get. But also, our ancestors gave up depending on the abundant fruits and instead moved into a novel landscape. It was a landscape with different kinds of foods—foods no other plant eater easily had access to, such as tubers and other roots.

Our early ancestors, Lucy and her ilk, probably had dug for tubers and other roots using simple tools, such as a sturdy stick. It seems unlikely that meat featured much in their diet, as the tools needed for hunting had not yet been invented. Tubers and swollen roots are an excellent food source, much denser in nutrients than leaves, but more difficult to find and harvest. Once dug up, though, they are plentiful, more than a single individual can eat.

In modern hunter-gatherer societies, the females typically forage for plant-based foods, whereas the males provide the protein via hunting and fishing. The standard narrative explaining our evolutionary history then goes something like this. Once the tools were invented to hunt large prey, the males went out together and later shared the proceeds of the hunt equally among themselves and the women and children. The women and children, in the meantime, had gathered plant-based foods, such as tubers, fruits, berries, and nuts. The need to get meat, for which one has to cooperate, then drove our early ancestors' social relationships. Hunting does become more efficient if the focus is on the hunt, and not on wasting time fighting fellow hunters. And so, the story continues, our ancestors became less aggressive to each other and better cooperators. It's a nice, simple story, but no, it's unlikely to be true.

For starters, the strict division of labor between men and women I started this discussion with is a gross simplification. Women often participated in the hunt, although their roles were different from those of the men. A more likely explanation for why we became so docile takes us back to before hunting even became possible. In that story, males and females stuck together to jointly care for their young in an environment full of predators. As our early two-legged ancestors adjusted to their new environment, parents sticking together became a necessity for the reasons I have explained. And that created the perfect conditions for females to share the food they harvested with their long-term mates, who in turn defended them and their children. That, at least, is the conclusion drawn from a recent study. I think it is a reasonable conclusion, for the simple reason that we know that by becoming bipedal, our ancestors became extremely vulnerable, needing to band together, and their new environment was jam-packed with new food sources high in nutrients. These new food sources could easily be shared, but they were also easier to extract with more hands to do the work. The more effort put into the collection, the higher the rewards. Yes, meat played an important role in our evolutionary history, but not quite the way it is often portrayed. Meat allowed for an initial increase in brain size, in *Homo erectus*, but the need to stick together in a new and hostile environment was the first step that led to sophisticated cooperation among our ancestors. It was cooperation driven by the females and their dependent babies, not by hunter males.

WHO CARES?

Human babies are not the only ones who need to be cared for. But parental care is not universal, either. A large group of animals doesn't seem to do the baby-care thing at all, and that puzzled me for a long time. I decided to look more closely into when youngsters need care and who is most likely to do the caring. That may seem a distraction from the main point of this book, but as an evolutionary biologist I think it is important to look beyond "just" humans to understand why we became the way we are. So please indulge me.

The non-carers just mentioned are in fact a large number of animals: almost all reptiles and amphibians and many species of fish. The puzzle I wanted to solve was why so few species in those groups, all cold-blooded vertebrates, provide for their offspring. Many of them guard their eggs or their young, protecting them against intruders and potential predators, but hardly any cold-blooded vertebrate parent provides its young with food. A look at the few exceptional species in which a parent does provide nourishment turns out to be illuminating.

Frogs that lay their eggs in phytotelmata (aquatic microhabitats in leaf axils, flowers, tree holes, bamboo stumps, and nut capsules) condemn themselves to having to feed their offspring, because the phytotelmata contain no nutrients for the growing tadpoles. The males nudge the females to lay infertile eggs—trophic eggs—in the water on which the tadpoles feed. Here's another example. This time, mom doesn't put enough resources into the egg so the offspring emerge completely helpless. Mom must now come up with plan B because otherwise the young will starve. This mom is the female of a weird amphibian that looks like a tiny worm but isn't, a so-called caecilian. To make up for her penny-pinching when it comes to eggs, the caecilian mom now lets her offspring eat her skin until they are big enough to venture out themselves. I know what I would do if I had the choice of putting a little more effort into an egg or letting my kids nibble bits off me. Another species of caecilian found a better solution, one we know quite well: produce milk for the mini-caecilians. Yup, milk-producing worm-lookalike amphibians do exist.

The last examples I found of parents involved in their youngsters' feeding are some species of cichlid fish. Fish fry are often tiny, many orders of magnitude smaller than their parents. Perfect little snacks for predatory fish. The larger the fry, the less vulnerable they are to being eaten. And so cichlids who live in habitats where predators are common assist their offspring to grow as fast as possible by providing them with food. Some cichlid parents ingest food from the bottom of the lake and expel it in mid-water so the fry can feed from it. Others allow their offspring to nibble from the mucus they produce externally. Like the caecilian.

This handful of examples of provisioning parents in a vast number of cold-blooded vertebrate species makes a stark contrast with the warm-blooded vertebrates—birds and mammals. The majority of bird species have altricial young that need to be fed prodigious amounts of food, especially those with a large brain. And all mammalian mothers spend a big chunk of their energy producing milk. Something makes cold-blooded vertebrates fundamentally different from all other vertebrates. What is it that lets most reptilian, amphibian, and fish parents off the hook?

It turns out the answer is deceptively simple. Cold-blooded vertebrates have a physiology fundamentally different from that of warm-blooded vertebrates. While warm-blooded vertebrates need a high energy intake to maintain a stable body temperature, cold-blooded vertebrates don't. And that means that reptilian, amphibian, and fish young generally don't need their parent's help obtaining enough nutrients. Unless there are some weird circumstances, such as in the examples I have mentioned, parents are better off eating for themselves and ignoring their young.

The physiology of cold-blooded vertebrates also explains why my pet snake, an Australian carpet python, refused her food in winter. With ambient temperatures low, she simply slowed down her metabolism and did even less than she did in summer, which wasn't a lot to begin with. (Snakes are not the most engaging pets, I came to realize.) Almost all bird young and all mammalian young are so small at birth or hatching that they risk dying of heat loss, so the challenge for the parent(s) is to make them grow as fast as possible—and that costs a lot of energy. The young of our cold-blooded counterparts have no such constraints, as they are not even trying to keep a constant internal temperature. Their physiology does come at a cost, though. Did you ever encounter a large-brained reptile, amphibian, or fish? I didn't think so.

One part of my original question has been answered. Parents provide for their offspring if the offspring otherwise have a high chance of dying. Now, what about the second part of the question: Who does the caring?

Fish are exceptional because there are many species in which the father is the sole caregiver. He guards the eggs and fry, fans the eggs if necessary, and removes parasites if needed. In these species, males hold territories to which they attract females, or they build a nest in which females lay eggs, or both. Here is the key: The females in those species with exclusive male care are egg layers (as opposed to live-bearers). Imagine a fish courtship. Let's say our male is a species of pufferfish that lives off the coast of Japan. These puffers construct an exquisitely beautiful nest resembling a raked Japanese garden, all out of sand on the bottom of the ocean. As he swims along the seabed, he uses the movement of his fins to rearrange the sand, creating a stunningly geometric structure. Sand particles not in the correct place or other eyesores are picked up and moved. The male spends an inordinate time constructing and maintaining his nest—because perfection is what his potential mates are looking for. Only the most perfect nest will satisfy a female pufferfish. Not that we know exactly what it is she is looking for, but the male seems to know, and if he does it right, an interested female will join him above his splendid arena.

Once there, the pufferfish don't actually mate. When they have decided "we're on," both male and female release their gametes. The female goes first. She releases her eggs and promptly does a runner. The male then releases his sperm over the eggs as they gently descend to the sandy bottom. The male is now stuck with the eggs, forced to look after them to make sure they don't get eaten, washed away, or parasitized. The male has no choice. Because sperm is much lighter than eggs, he cannot release his sperm before the female has spawned, as the sperm would drift away before there are any eggs to fertilize.

Most bird parents do it together, feeding the young and keeping them warm. They have to, as the little ones are so small they are unable to maintain their own body temperature unless they receive a lot of food—more than one parent can provide. Eating loads of food makes them grow fast so they quickly get out of the "thermal danger zone," when they are too small to prevent excessive heat loss. All small-bodied warm-blooded animals have an unfortunate surface-to-volume ratio. That is, too much surface through which heat escapes and too little volume to generate heat. Being small-volume myself, I can attest to the veracity of this argument. There is a good reason I moved to the tropics.

Mammalian young suffer the same thermal constraints as do birds, but you'd be hard pressed to find a species in which the father feeds the young as much as the mother does. He simply can't, even if he wanted to, as he doesn't produce milk. Both bird parents, in contrast, are equally capable of providing their young with food, and so in most bird species (a whopping 90 percent), we see biparental care. For the same reason, every grown-up can feed a baby bird; alloparenting—care by individuals other than the parents—is also quite common in birds. Alloparents are often relatives of the breeding pair—young from previous breeding seasons. And not all of them stay at home to help by choice.

White-fronted bee-eater males are often forced by their fathers to give up their own attempts at nesting and instead move back in with mom and dad to help rear siblings. A father's harassment is most successful when the son is in his first or second year of attempting to do his own thing. Why would fathers harass sons and why would sons let dad be successful? It's easy to understand dad's motives; he gets help (rather unlike human dads, I'd guess), but what is in it for the son? Raising relatives, most likely. Successfully establishing a nest and raising a clutch of offspring is hard, and young inexperienced parents often fail. By helping dad (and mom) raise siblings, a young male can still gain indirect reproductive success by helping his parents. If both his parents are still breeding together, then the young bird's relatedness to siblings is identical to his relatedness to his own would-be offspring. So, in genetic terms, he doesn't lose any fitness. If his dad re-partnered, he would be raising half-siblings, not as good as full siblings, but still much better than nothing if he is unsuccessful doing it himself. By raising siblings, the young bird gains valuable experience, so his own attempt to independent breeding next season might have a better chance of success. (You may wonder why daughters are not harassed. In this species, the daughters move away from the parental territory so they are not within their parents' reach.)

That leaves us with mammals. In 90 percent of all species, it is mom, and mom alone, who looks after the young. There are two reasons for father's lack of interest. Or apparent lack of interest. One was already hinted at: Fathers can't produce milk and all mammalian young depend exclusively on milk for a while. But what's the second reason? Males can literally get away with abandoning the female. She is the one who is pregnant, for weeks or months, during which there isn't much the dad can do to help grow his offspring. So, like the Japanese pufferfish female, off he goes, doing other important things.

At least that used to be the explanation for the lack of paternal care in mammals. Dad can't do much to assist mom and is therefore more likely to pursue other females. But as so often, the truth is much more interesting. And varied. A few mammalian fathers do actually stick around—roughly 10 percent of all species. Here is the thing. In general, there are equal numbers of males and females, and it takes two to tango so every offspring has two parents. While it is true that some males can monopolize more than one female depending on the mating system—think chimpanzee harems—such conditions are rare. So, what is a male going to do while "his" female is pregnant or nursing? If he does not have any other mating opportunities, he might as well stay around and help out. He can provide the female with food, defend her and the offspring, or just carry the little one around so mom is unburdened for a while. Mom can now increase her own food intake, which means she can produce more milk, and the young will grow faster and wean at a younger age. Both parents benefit because the next young can now be produced sooner.

Like the birds, some mammal species even enlist relatives to help out, turning mom into a breeding machine. The naked mole rat is such a species. I think it qualifies as the ugliest mammal on earth. The dominant female is basically a pup and milk producer while her offspring, both sons and daughters, do all the work. They live underground in the deserts of southern Africa, feeding on tubers, which as we've seen earlier are hard to find but are a bonanza when found. Their biology has often been compared with that of a bee colony; not an unreasonable comparison, although bees are much nicer to look at. But there is another super-social mammal. No, not humans. Meerkats.

ALL IN THE FAMILY

Years ago, while doing fieldwork in South Africa, my husband and I decided it was time for a break. And so we traveled north, to the Kalahari Desert, to visit a research station that became famous thanks to a successful television series. One doesn't normally associate a biological field station with a TV documentary series, as field stations tend to be messy, grotty, and full of eccentric biologists. But this one had a famous family living on its premises. The Whiskers: the family of meerkats whose dynasty tales became *Meerkat Manor*.

The success of *Meerkat Manor* and the money that flowed from it has secured the ongoing research on meerkats and naked mole rats at the Kalahari Desert Research Station. (The naked mole rats did not feature on TV, for obvious reasons.) For decades, researchers and volunteers have worked to habituate the meerkats to human presence from the minute they leave their burrows in the morning until the moment they returned to go to sleep. The result has been amazing. Imagine this. A meerkat sleepily comes out of the family burrow just after sunrise, when the desert air still has some coolness. Waiting for him or her is a trusty volunteer with a scale (plus occasionally two Australian visitors) and a container with some boiled egg. Gingerly the meerkat steps onto the scale, still not quite awake. The volunteer writes down its weight and rewards the meerkat with a small piece of egg. She knows every individual by name. She or another volunteer will weigh the meerkat again before it retires to the burrow in the evening. During the day, all the meerkat group's activities are recorded. What they eat, whom they interact with, and which pups they babysit. Because that is what meerkats do: they look after the pups collectively, teaching them how to safely eat venomous scorpions (considered a delicacy), and even providing them with milk (allolactation). The latter process is particularly interesting, as the dominant female does not tolerate other females that are pregnant, nor their young. Females who do not obey the social order are forcefully evicted by the dominant female, and if they gave birth, their young are murdered. The cruelty of the superficially adorable critters was driven home by the main researcher, Tim Clutton-Brock, during a talk he gave to a group of animal behavior researchers, which I attended. One photo in particular has stuck in my mind: a decapitated meerkat pup. Its crime? Being born to the wrong mother.

In providing milk to the reigning female's pups, the other females may try to hide their own pregnancies. Others may nurse the young as payment for their disobedience, when they return to the group after having been evicted. The dominant female's iron fist allows her to produce many more young than she could possibly produce on her own. The other group members reap indirect benefits, as they are all family.

We are quite like meerkats. Without others' help we couldn't produce the number of babies we do. Although we are nicer to other pregnant females.

Ever since our ancestors became bipedal and lost most of their hair,* babies have had to be carried around in moms' arms. And that means that mom couldn't do much else unless she could either share the carrying with others or leave the baby in a safe place. She probably did both. But no one asked the babies. . . . Human babies—and this must also have been true for our ancestors' babies—need the security of being cared for as much as any other dependent young. And like other dependent young, human babies have a bag full of tricks for exploiting the innate tendency of older individuals to care for them. I am not talking about maternal instinct—a myth—but the instinctive "need" of both sexes, young and old, to protect those who are incapable of doing so themselves.

SURVIVAL OF THE CUTEST

One time I could not give a scheduled lecture on the conflict between parents and their offspring, for reasons that now have escaped me. Luckily, a younger colleague who had just returned from maternity leave was happy to gain more teaching experience. And so I handed her my PowerPoint presentation, told her to make any changes as she saw fit, and went to do whatever it was I was doing instead of lecturing.

One of my slides showed a collection of young mammals, each extremely cute. Even now, when I look at that slide I just want to pick them all up for a cuddle. But something had

---

* In truth, we didn't lose our body hair, but our hair changed so it is now microscopic and useless for holding on to.

changed. In the middle row there now appeared a baby human among the adorable animals. No prizes for guessing whose baby that was.

The baby collage perfectly illustrates an interesting point. Young mammals, almost without exception, are adorable. Not just to their parents but also to everyone else. And it is not just humans that find baby mammals adorable; in her book mentioned earlier, with the thought experiment of chimpanzees on a plane ride, Sarah Hrdy tells the story of a lioness nicknamed Kamuniak, who adopted five baby antelopes. One she ate, but the other four she seemed to care for as if they were her own. In the end, this adoption didn't quite work out well for the little fawns, but it illustrates how strong the attraction of immature, vulnerable creatures can be, even if they are something one normally eats.

I found a similar story in a blog, which I used for one of my lectures. This time, a baboon baby convinced the lioness to protect it after that same lioness had killed its mother. The baby was later snatched from the lioness by a male baboon, presumably the father, who had been watching the ordeal from a distance. Hrdy refers to babies as "sensory traps," having evolved characteristics that make them irresistible. Who hasn't been at least tempted to take home a kitten or puppy that needed a new home? (If you were wondering how I obtained my dog, now you know.)

When you are vulnerable and depend completely on others for your needs, your best bet is to become irresistible, evolving features that play to your potential carer's senses. The famous ethologist Konrad Lorenz, who shared the Nobel Prize with Niko Tinbergen and Karl von Frisch, called the specific features that young animals, including humans, have evolved *Kindchenschema*, or "child cuteness." A large head relative to body, large eyes, soft-elastic surface texture, and round and protruding cheeks all combine to elicit caregiving behavior in the observer. But sounds and smells also play a role. Ever smelled the head of a baby?

Human babies are different from goslings. A gosling will attach itself to the first figure it sees when it hatches from the egg. Goslings are precocious young birds, but they need to be protected from predators and apparently anything alive will do. Even a researcher, as Lorenz found out. The photo of Lorenz walking through a field followed by a string of goslings would have become an internet sensation, had the internet existed at the time. The little birds imprinted onto the Nobel laureate, as he was the first living thing they saw.

Our babies are not like goslings, because they imprint on any human who cares for them. For the first twelve months or so, babies tend to greet everyone with a smile and adorable coos, even complete strangers. When they are about a year old, they have figured out that there are some "special" people—people who treat them particularly well. For a while, they become attached to their main caregivers, not just mom and dad, but also everyone else who has played an important role in the baby's first year. Now, strangers are scary. They are no longer needed, as the baby has secured its complement of caretakers.

Because of their utter helplessness, a baby's first priority is to ensure continued care. But there is more. Building healthy social relationships immediately after birth is also essential for normal brain development, and as we'll see later, for the development of language. The importance of early relationships sadly became clear from the horrors of Romanian orphanages.

Romania's former dictator Nicolae Ceaușescu forbade any form of contraception, and so the poorest women were often forced to abandon their children in an orphanage. In total, some 170,000 children were raised under the most deprived conditions. (These were the ones discovered after the dictator's fall in 1989.) They were not physically harmed, but were emotionally neglected. Except when they were being fed, bathed, or diapered, the babies were left in their cribs with no mental stimulation and no ability to form any sort of relationship with another human being. Upsetting as their plight was, it provided a unique opportunity to study the effect of neglect on the normal development of young children, as all the babies had been delivered to the orphanage shortly after birth.

A total of 136 children, ranging in age from six months to three years, were assessed by a trio of psychologists and became part of a long-term study. After the initial assessment, half the children were placed in foster care, while the other half remained in institutional care. Over subsequent months and years, the children continued to be assessed, in addition to a control group of children who had never been institutionalized.

The children born into neglect had many problems. Their cognitive function, language skills, and motor development were all delayed. They also had smaller brains. Those who were placed in foster homes before they turned age two caught up somewhat, but never completely. One characteristic that never quite caught up was their brain size.

One of the study's main conclusions was that the first two years of human life are essential for healthy brain development. All young mammals have a sensitive period during which the brain develops normally, given the correct conditions. But our babies' brains are extra sensitive because they are so undeveloped at birth. Provided they are lovingly cared for, this immaturity turns them into extraordinary learning machines.

Learning is what childhood is all about. Or, as American professor of psychology Alison Gopnik puts it, childhood is *for* learning. Gopnik compares babies to a research and development department in a major corporation, where the wildest theories are tested and the ones that turn out to work add to knowledge about the world (at least that is how R&D departments should function). Young children do not yet have preconceived ideas about how the world should work—they must figure it out as they go along. They explore. They try stuff out. And to do that they need a safe environment, with responsible others around them to keep them alive. As with any research and development department, children need complete freedom to come up with the best inventions. And they need excellent support.

Jamie, our New Caledonian crow mentioned in Chapter Two, spent her childhood learning how to make tools. For the first two years of her life, Jamie would have starved to death if she had to depend on her own tools. Lucky for her, her parents made sure she got enough to eat while she explored different ways to construct the implements she needed. Human babies take much longer to become independent than Jamie. A hunter-gatherer youngster won't bring in the amount of food she needs to thrive until she is about fifteen years old. During that time, there is a lot she needs to learn. And not just from mom, as mom can't raise her on her own, especially when there are siblings. Mom needs help. Luckily, help is nearby, because with a long childhood comes a long life. And that means there are some individuals perfectly primed to help. And to learn from. Grandparents. Particularly grandmothers.

Most animals don't live long enough to see their children's children. We do, because a large brain takes a very long time to grow.

BREAKING THROUGH THE GRAY CEILING

I now live in the land of koalas. While almost every visitor to Australia probably has seeing koalas on their wish list, the iconic animals are actually rather boring. They spend twenty hours a day asleep, curled up in the fork of a eucalyptus tree, the leaves of which are their only food. Their diet is so poor and so full of plant toxins that it takes a very long time to digest the food. A poor diet means a slow metabolism, and spending the majority of your life asleep seems a good solution.

Koalas may be boring, but at least they are not as pathetic as giant pandas. Pandas are actually bears, and bears are normally carnivores. But not pandas, which, like koalas, eat only one type of food. At some stage in their evolutionary history, pandas started to rely on a bulky, fibrous bamboo diet to which their (carnivore) gut was not adapted. The gut of an herbivore has special adaptations to assist in the digestion of plant materials—the cecum being one, which in koalas contains a large number of specially adapted bacteria that detoxify the eucalyptus leaves—but the panda's gut has none of these. To compensate for the lack of nutrients, pandas started to increase in size, because an increase in size lowers the metabolic cost per unit of mass in organisms that maintain a stable core temperature (homeotherms, or warm-blooded animals). A larger body also needs less protein per unit mass. At some stage, giant pandas became so large they couldn't even hunt if they wanted to, thereby condemning themselves to the life of a rather pathetic bear.

You may wonder why I thought it necessary to impugn the reputations of two iconic animals, but both illustrate how diet affects the raison d'être of an organism. If your normal food is low in nutrients and hard to digest, you are compelled to dial down your metabolism (koala). If you rely on eating something to which your physiology is not adapted, you compensate by becoming larger so you require fewer nutrients (panda). We humans evolved from an herbivorous ape and have the physiology to prove it. Our digestive system is similar to that of our closest living relatives, structurally but also functionally. We all have a simple acid stomach, a small intestine, a small cecum, and a segmented intestine that increased the capacity to extract nutrients before waste is excreted. And we take a long time to pass food through the whole system, so the most nutrients can be recovered. (Yes, some biologists make a career of measuring the time it takes between eating and pooping across a variety of animals. Such a person discovered, for example, that the southern cassowary goes from swallow to poop in as little as three hours.)

As the giant panda shows, one strategy to subsist on a low-quality diet is to grow big, a path that gorillas and orangutans appear to have followed. Their large body size allows them to survive on a diet of leaves, bark, and unripe fruits. That's a good strategy if most of the food around you is of low quality, but the adaptation comes with a price. They became a bit like the koalas. The low nutrition in their diet does not provide them with much excess energy—energy that could be spent on moving around more. Nor do they have the energy to grow a larger brain.

Our early ancestors took another route altogether. By moving out of the trees and into a new environment, they gained access to plenty of good forage, even before the addition of meat to their diet. Once they lived in groups to overcome the downsides of walking on two legs, getting that good food became even easier. Brain size started to increase slowly, reaching a temporary maximum in *Homo erectus*. But something prevented a further increase in brain size. *Homo erectus* had reached the so-called "gray ceiling."

Here is the twist. A large brain takes a long time to grow. The larger the brain, the slower the animal develops. And the slower it develops, the later in life it will start to produce its own offspring. The danger, then, is that one doesn't live long enough to get to the age at which one can reproduce. A species that reproduces slowly is extremely vulnerable to freak environmental events. Think droughts, flood, fires—you name it. Such freak events can kill off individuals before they have had a chance to make offspring. And that means that slow-maturing species are prone to extinction. At some stage, the risk of dying before having produced babies becomes so great that life span can no longer increase to allow a larger brain.

Our large-brained relatives could not further increase their brain size, as that would have required producing more babies to offset the increased risk of dying before having had the chance of becoming a parent. What made our species special so we could break through the gray ceiling? We asked for help.

Just like meerkats and naked mole rats, we became cooperative breeders. But cooperative breeders with a twist. Instead of enlisting our children to help raise more children, we enlisted our parents. In particular, our mothers. We could, and can, enlist our parents because after the initial increase in life span associated with an increase in brain size, life span increased even further. Some women started to live decades beyond their reproductive age, a phenomenon unique among primates. But that "invention" appears to be relatively new, evolutionarily speaking. The teeth tell the tale.

Assessing the wear on teeth is a relatively simple method to guess the age of an individual. If you then have teeth from many individuals you know lived in the same place at the same time, you can estimate how many young people there were for every older person. Such data suggest that extreme longevity—the number of individuals surviving to older adulthood—evolved fairly recently. Longevity started to increase, slowly and steadily, from the Australopithecines onward; at the same time, brain size also increased, slowly and steadily. For every Australopithecine young adult, there were, on average, 0.12 older adults. In early *Homo,* that number had doubled to 0.25, reaching a temporary peak of 0.39 in Neanderthals. But then, about 50,000 years ago, that number skyrocketed to 2.08 in *Homo sapiens*. For every youngster there were now two adults. That ratio is a massive difference between our species and that of the Neanderthals. This shift in ratio also seems to coincide with other changes in human behavior, particularly the invention of more sophisticated technologies that seem to require some form of cultural transmission. Coincidence? I don't think so.

While other cooperative breeders make use of an additional labor force in the form of younger individuals to help raise offspring, humans made use of the accumulated knowledge that resides in older individuals. Our species managed to spread around the globe, feeding on the highest, most nutrient-dense and difficult-to-acquire plant and animal foods. Finding and extracting such foods requires brain-based skills—skills that take a long time to accumulate. But once they have accumulated, they can be transferred, provided there is an effective means to do so. And what could be more effective than language?

Here is our origin story, as I see it.

Let me take you back to 1.9 million years ago, to the fossil of a young female with the exciting name KNM-ER 2596. We don't know which species she belonged to, because multiple human-like creatures lived in the area at that time. We do know that KNM-ER 2596 suffered some misfortune, as she had broken her ankle. The break didn't cause her death, as the break healed. Think about that for a second. Here we have a bipedal creature, living in an environment surrounded by predators. And now one of them can't even walk, let alone run, to save herself. KNM-ER 2596 must have been part of a small group, probably comprising family members, who went through the bother of protecting her, looking after her, until she was well enough again to walk. Our ancestors had to become social to stand any chance of surviving out of the trees, on two legs. Even without a broken ankle.

The social relationships our ancestors formed did more than protect them from predators. It became easier to find and collect food and to look after babies and toddlers. Better food made bigger brains, which made for more complex social relationships. When, owing to a fluke of nature, first the repair and then the duplication of a brain-growing gene, our species' brain started to grow even more; all the requirements were in place to make sure the now seriously underbaked youngsters could thrive and survive.

To care for such needy creatures requires a high level of cognitive skills. The more immature the young, the more skills are required to raise them to adulthood. As our babies became more helpless, they also became better learners. Their neuroplasticity allowed them to soak up information from their social environment, accelerating the evolution of even higher cognition. The more helpless our babies became, the more they depended on others to look after them, shaping those babies into the tiny manipulators they are today.

Underbaked babies, on the one hand, and knowledgeable elders, on the other. Bigger brains, longer life. A longer life provides people old enough to become repositories of information. Grandmothers started to contribute to the "embodied capital" of grandchildren. When it takes an inordinate length of time for children to become independent, there comes an age when a female is better off helping raise grandchildren than more of her own children. Menopause became a stage of life. The knowledge and experience embedded in grandmothers could now be funneled into raising their daughters and sons' children.

When our species' head was remodeled to accommodate a ballooning brain, we made the best of the design fault that came with our changed throat. Our human-specific brain fluke gave us the best tool to perfect our social life and our childcaring skills. Mom could now ask for help. Granny became an effective teacher. And our babies? They come into this world with a unique skill: the ability to learn language.

six

WHO NEEDS HALF A GRAMMAR?

Despite being the co-discoverer of evolution by natural selection, Alfred Russel Wallace struggled with the idea that language evolved. He didn't see how a little bit of proto-language could be useful, going so far as to argue that there is no problem we can solve using language that we can't solve just as well without it. Explaining how and why language arose, therefore, came to be called "Wallace's problem." Part of the problem was political, given the role language plays, then and now, in carving out a special spot for humans in the tree of life. The other part of the problem was that it seemed impossible to come up with testable hypotheses about the origin of language. To avoid a string of unconvincing papers, both the Société de Linguistique de Paris and the London Philological Society *banned* the discussion of the evolution of language in 1866 and 1872, respectively.

The ban was broken in the late 1950s, first by the psychologist and behaviorist Burrhus Frederic Skinner and then by the linguist and philosopher Avram Noam Chomsky. Chomsky challenged Skinner's claim that language is just a special form of conditioned behavior—a type of associative learning, familiar from simpler organisms. "Learning" language isn't a matching game, Chomsky argued, but is, rather, the activation through experience of an innate faculty unique to humans. Only humans have a "language acquisition device," a feature that somehow came into existence during our evolutionary history, presumably due to a mutation. So, Chomsky actually agreed with Wallace, in a way. The gulf between human language and other forms of communication in the animal kingdom seemed to need a bridge too far to cross by a slow accumulation of minor changes. Half a system of grammar isn't better than no grammar at all. A genetic fluke seemed like the only answer.

Skinner and Chomsky had diametrically opposed views on the function of language, and therefore its origin. According to Skinner, language serves as a means of communication, to share knowledge. In Chomsky's view, language helps us think. Unfortunately, these hypotheses for the function of human language don't help much in solving the problem of its origin. Learning to associate a word with its meaning by trial and error (Skinner's view) does not explain why young children grasp the grammatical rules (syntax) of their language in the first few years of their life. Chomsky's idea of an innate ability to understand grammar does not explain the origin of words, and without words, there is no need for grammar. Perhaps because of his prominent place in the world of linguistics and political economy, Chomsky's views on the evolution of language garnered widespread acceptance among linguists and the general public, despite a complete lack of evidence for the language acquisition device, or even a hypothesis for how the device came about. But *is* language uniquely human?

On July 8, 2011, the U.S. production company and film distributor Roadside Attractions released the documentary *Project Nim*. This documentary details the tragic life of Nim Chimpsky, a chimpanzee who was stolen from his captive mother at the age of two weeks and raised in a human family along with a human child of similar age. Nim's name was an allusion to the aim of the project: to prove Noam Chomsky wrong by showing that nonhuman apes can learn language under the right circumstances—in this case, by being raised from infancy to communicate with American Sign Language (ASL).

After a couple of years, the researchers separated Nim from the family because he'd grown possessive of his adoptive mother and jealous of his adopted father. His tuition continued, though, with a string of young female teachers, until Nim flew into a rage, seriously wounding one of his teachers. The experiment was aborted and Nim was moved to the Institute for Primate Studies in Oklahoma, where he struggled with being a chimpanzee. He'd never met another nonhuman ape. Nim identified as human.

The project failed to refute Chomsky. It did show that chimpanzees can learn to associate signs with objects. But associative learning does not require grasping the concept of the object or the meaning of the symbol; it can merely involve internalizing a routine for turning prompts into treats—rewards for indicating or retrieving the intended object. Impressive? Sure, but it's not a talent unique to apes. Separate studies taught dogs to fetch objects identified by their English names. One dog mastered more than 200 words, another 1,022 words. In contrast to the ape studies, no one expected dogs to have language, so the results were reported "simply" as associative learning. I use quotation marks for *simply* here, as I think the dogs' achievements are far from simple. Had they been apes, their achievements would most likely have been interpreted very differently. Bees, too, can readily learn to make associations between prompts and food rewards. But, again, no one would argue they have the ability to use human-like language. They learn to make associations because that makes them better at getting food. So did Nim. He quickly figured out that if he played the game, he would get something yummy.

In the end, both Skinner and Chomsky were half right. Language is partly acquired and partly innate. Full-blown linguistic competence takes associative learning to amass a vocabulary and a native faculty for universal grammar. So, was there a language mutation in our evolutionary history that would solve Wallace's problem? For a while several influential people thought there was.

THE LANGUAGE GENE

In 1990, a UK family found itself in a genetics clinic after seven of its children had attended the same speech and language unit. There was something odd about the seven children. Even though they were of normal intelligence, they were unable to speak properly. They had normal control over their facial muscles, tongue, and palate, yet when they tried to say something, gobbledygook came out. For some reason, they could not control their mouth, lips, jaw, and tongue to make normal sounds.

They knew what they wanted to say, but they couldn't make the words nor construct complicated sentences. They suffered from apraxia.

The children's communication problems were unique, known at the time only from the one family. Surely their problems were due to some sort of genetic issue. Based on the not unreasonable hunch the family's speech and language problems had a genetic basis, the whole family was referred to the genetics clinic.

In total, sixteen family members across three generations had severe speech problems. That meant more than half the family was affected. When one parent had speech difficulties, many of their children did, too. The one parent in the family who did not suffer from the condition had children without speech problems. So, the way in which the condition was inherited pointed to a single, genetically dominant gene as the culprit. Perhaps the sufferers carried a mutation in a gene that was necessary for language.

The language gene! There were additional reasons to think the language gene was about to be identified. Affected family members also had problems identifying basic speech sounds, understanding sentences, judging what was grammatically correct, and what not. They also failed to come up with the past tense of invented verbs—something small children have no difficulty doing.

Then, in 1998, a group of researchers located a small segment of chromosome 7 that seemed to be correlated with the family's speech disorder. They were careful not to refer to this locus, *SPCH1,* as *the* language gene, but merely as evidence for a genetic component of the ability to understand grammar and for the development of speech. That all changed when the cause of the speech disorder was further narrowed to a single gene, *FOXP2*. We now know that *FOXP2* is essential for the development of structures in the brain that are important for speech and the ability to understand language. It seemed that a single mutation in this gene messes that all up. At the time, a conclusion was easily drawn: *FOXP2* is a viable candidate for being the "language gene."

The plot thickened when, in 2002, a team of researchers from the Max Planck Institute for Evolutionary Anthropology, in Leipzig, Germany, and the Wellcome Trust Centre for Human Genetics, at Oxford, discovered that the human version of the *FOXP2* gene carries a distinctive mutation. The gene itself is not uniquely human. In fact, it is an old gene, found in rodents, birds, reptiles, and fish, among others. But our version of *FOXP2* appears to be special. It differs from that of the other great apes in two changes that lead to different amino acids in the FOXP2 protein. In the other apes, positions 303 and 325 within the *FOXP2* gene carry the code for the amino acid threonine and asparagine, respectively. But in our version, those positions instead code for asparagine and serine. Minor as these changes may seem, they drastically change the human FOXP2 protein relative to the version found in the other great apes.

Even more intriguing is that the family members with speech and language disorders all carry a faulty copy of *FOXP2*. Their copy has a change in one letter of the genetic code that renders the FOXP2 protein it codes for useless. All affected family members carried one normal version of *FOXP2* and one faulty version. For normal speech development, one needs both *FOXP2* copies to be functional, both coding for the normal protein.

When one of the copies doesn't function, only half the required amount of protein is formed. And that isn't enough. We learned that fact through some clever experiments on birds and mice.

One flow-on effect of the Human Genome Project was the development of molecular toolkits that facilitate no end of genetic wizardry. Using some of that wizardry, researchers manipulated mice and zebra finches in such a way that they, too, produce only half the normal amount of FOXP2 protein. Zebra finches, being songbirds, learn their species-specific song from tutors—older birds in their neighborhood. But birds with half the normal quotient of FOXP2 protein can no longer accurately mimic the songs of their tutors. It seems that, in the absence of the full complement of FOXP2 protein, the birds became bad at learning songs. Mice, too, are negatively affected when the expression of *FOXP2* is dialed down to half the normal level. Normally, mouse pups produce ultrasonic calls when isolated from their mother, but not those with the manipulated *FOXP2* gene. They went quiet.

The *FOXP2* gene is expressed in many tissues, lungs, heart, and even bone, but it is the brain that relies on proper functioning of the gene. Or, to be more precise, the brain needs the proper amount of the protein that the gene codes for. Somehow, the FOXP2 proteins affect the development and function of neuronal circuits important for vocal skills and control of speech-related muscles. That's true in humans, but also in other animals.

Using more genetic wizardry, a large team of researchers led by Christiane Schreiweis introduced two copies of the human version of *FOXP2* into mice. The result? The humanized mice became better learners, thanks to changes to their brain.

The last piece of evidence for *FOXP2* as the language gene is *positive selection*. Here is the idea. Sometime in the last 6 million years or so, after the split from the common ancestor with chimpanzees, a mutation in *FOXP2* caused some of our ancestors to become better at vocalizing. They became better communicators. Because communication had many other benefits, the mutation started to spread, until every human now carries it—bar the UK family with the mutation. As population geneticists express it, the human version of *FOXP2* had "gone to fixation." Every human alive today carries the same version of the gene. We became the talkative ape courtesy of the language gene—the human-specific mutated version of *FOXP2*.

Convinced? Many were, until 2018, when this story started to unravel.

A large U.S. team led by Elizabeth Atkinson shed doubt on the idea that the human version of *FOXP2*—the gene with that specific mutation—has been specifically selected for in our species. To stand any chance of being *the* language gene, the mutation that allows us to speak should have occurred 100 to 200 thousand years ago, early in our evolutionary history. But Atkinson and her team found it is far more likely the human *FOXP2* mutation occurred much earlier, before our ancestor split from the lineage that led to the Neanderthals and Denisovans—for they, too, have the mutation. We don't know for certain Neanderthals and Denisovans didn't speak, but there is no persuasive evidence they did. In fact, as I will argue later, there is pretty good evidence *against* speaking Neanderthals (we don't know much about Denisovans because we don't know what they looked like). Therefore, Atkinson's finding is compelling. If Neanderthals (and Denisovans) had the same mutation as us, modern humans, then the human-specific *FOXP2* mutation cannot be the reason why we speak.

In a further assault on the language-gene hypothesis, Atkinson and her team repeated the analysis that had shown evidence of positive natural selection on *FOXP2* in *Homo sapiens*, but using a much larger sample of human genomes, from a much larger geographic range. That study failed to show that all humans share the same mutation. So, it now seems that while *FOXP2* is essential for human speech and helps other animals make sounds, it does not give language to any animal that carries it. A "language gene" it is not.

WHO NEEDS A MIRACLE?

For both Alfred Wallace and Noam Chomsky, language is so different from anything else in the natural world that its existence almost requires a miracle. To me, this argument is eerily similar to the argument creationists use to dismiss the evolution of complex organs like eyes via natural selection. The human eye is too complex to have evolved step by step, they argue, for what is the selective value of an eye before it can see? Nor is half a wing of any use. Therefore, creationists assert, both the eye and the wing must have been created by you-know-who.

I am not saying that Wallace and Chomsky are creationists, they are not. I am saying that their argument doesn't make sense. Even though no other animal uses words as we do, it is a gross simplification to assume that language started with syntax and words. Rather, it no doubt began with nonverbal communication that had some survival value but wasn't yet a language. But before I go into details, let's pause for a moment to think about

Chomsky's native faculty for universal grammar. His language-acquisition device is unique, he argues, to humans.

In his 1994 book *The Language Instinct*, Steven Pinker starts his chapter 9 with the amusing real-life headline "Baby Born Talking: Describes Heaven." The headline was from the newspaper *The Sun*, published on May 21, 1985. Similar headlines appeared in later years, all reporting the births of miraculous talking babies. Just in case you were wondering, the other articles were not published in scientific papers either. All were published in *The Sun*, a UK paper not known for its scientific integrity. Of course, Pinker didn't use *The Sun*'s claims to argue that babies are born capable of speech; they aren't. But they are capable of quickly grasping the subtleties of the language they hear around them. They have a language instinct, a native faculty for universal grammar. They go from random babbling to speaking in full sentences in just a few years. And not just in one language. When exposed to multiple languages from a very young age, they easily pick them all up. I have friends whose children are completely bilingual, speaking English with their dad and Swedish with their mom. I've never heard the children mix the two languages; they simply swap effortlessly depending on whom they are talking with.

It is this seemingly miraculous ability of children to learn language that makes Chomsky doubt the power of natural selection to bring on a language. But an innate ability to learn the sounds of your own species is far from unique among animals. For example, male songbirds of many species use their songs to defend their territories and to attract and retain mates. A female selects her mate based in part on the quality of the male's songs. Females "know" what a proper song should sound like because they recognize their species-specific songs, probably because that is what they have been exposed to. (Even when still in the egg, the developing bird can hear, just like babies hear inside the womb.)

Take the regent honeyeater, a critically endangered songbird endemic to the southeastern coast of Australia. Sadly, the number of regent honeyeaters living in the wild is now so low that some young males no longer encounter enough older males to properly learn their species-specific songs. Because here is the thing. Like human babies, young songbirds pick up "language" from those around them. And like human babies, they use whatever they hear to develop their personal song repertoire. That's provided the sounds come from birds, of course, unless the bird belongs to one of the amazing species capable of convincingly mimicking any sound they hear, including ringtones, fire alarms, lawn mowers, and human speech. This includes species such as the Australian lyrebird or the North American mockingbird. To aid the regent honeyeater and prevent it from going extinct, Australia initiated a captive breeding program so zoo-bred birds could be released back into the wild.

Unexpectedly, while the birds bred happily in captivity, their subsequent release did nothing to halt the decline of the wild population. Why not? As it turns out, the males didn't sing the right song. Instead of growing up around males that sing the proper regent honeyeater song, the hand-reared zoo-raised birds were exposed to different sounds—sounds they incorporated into their "zoo-specific" regent honeyeater male song. When the captive-bred birds were released into the wild, the local females snubbed them because they didn't have the right accent. Only zoo-bred females would give the zoo-song singing males the time of day. To boost the wild population of regent honeyeaters, conservation biologists will need to teach captive-bred males the correct song by making them listen to recordings of their wild counterparts.

The regent honeyeater's fickle females remind me a bit of my eldest daughter when she was about twenty-one months old. Like any normal twenty-one-month-old baby, she was happily babbling sentences with anyone willing to listen, eager to join the conversation without necessarily making much sense. We then decided to take her and her little sister on holidays to Thailand. Weirdly, the normally happy bubbling became desperately unhappy. Being so young, she could not tell us what made her so unhappy, so we had no choice but grin and bear her deteriorating attitude toward our tropical holiday. What on earth was going on in her little head? The mystery unraveled when we met up with some friends from home halfway through our trip. As quickly as she had become unhappy, she returned to her happy self.

It took us a little while to figure out what had gone through her head. For her, a massive change had happened beginning when she set foot in Thailand. People were speaking gibberish! She had just started to join in conversations, but now most people around her no longer made any sense. When we met up with people she knew, who spoke a "normal" language, she must have realized the world as she knew it still existed, somewhere. She must have then decided to trust her parents to one day take her back to that world. For the rest of the holiday, she remained the unencumbered little girl we knew. And her little sister? She was too young to be confused by the change in language; to her, all languages were equally interesting.

Our babies and young children have a keen interest in language. But some nonhuman apes do, too. It's time to meet Kanzi.

THE MIRACLE APE

Kanzi is the adopted son of Matata, the matriarch of a group of bonobos that live at the Yerkes Field Station, at Emory University, in Atlanta, Georgia. From six months of age, Kanzi was exposed to his adoptive mom's training program to learn symbols. Caretakers would speak to Matata in English while pointing to lexigrams. Matata turned out to be useless, unable, or unmotivated—who knows?—to learn to associate the English words with the symbols on the lexigram. But not Kanzi. Despite not having shown any interest while Matata was being trained, when he was two years old, Kanzi showed he understood what was being said to him. Six months later, he spontaneously started to use lexigrams to "talk" with his caretakers. Kanzi seemed the perfect subject to study the linguistic abilities a nonhuman ape can achieve.

To avoid the pitfalls of previous attempts to teach language to nonhuman apes, which in the end were simply means to obtain treats by making associations, the team led by the psychologist and primatologist Sue Savage-Rumbaugh made a monumental decision. Instead of teaching Kanzi more associations between English words and symbols on the lexigram, they decided to treat him like any human child. And so, in 1985, at the age of five years, Kanzi was moved to the Language Research Center at Georgia State University. He was later relocated, along with his sister, Panbanisha, to the Great Ape Trust, in Des Moines, Iowa.

At both places, Kanzi's sole exposure to language was via caretakers talking to him while going about their business. There are wonderful videos available on the internet of Kanzi exploring the forest with Sue Savage-Rumbaugh, lighting a fire, and cooking a meal together.

The videos show Savage-Rumbaugh telling Kanzi to collect sticks to make a fire. When he hasn't collected enough sticks, she tells him he needs more. Kanzi breaks the sticks into the right lengths and piles them up. He finds matches in Savage-Rumbaugh's pocket, because she tells him where the matches are. Once the fire is lit, Kanzi roasts marshmallows.

When Savage-Rumbaugh and Kanzi cook together, Kanzi seems to understand what is being asked of him. Before they can boil the potatoes, the potatoes need to be washed. When he doesn't put enough water in the pot, he is told more water is needed.

Watching the videos makes you feel that Kanzi truly understands what is asked of him. But how can we be sure he really comprehends language and is not simply guessing? It turns out there is a relatively simple way to test for true comprehension. Select a string of words whose exact meaning depends on the word order. As Steven Pinker writes in *The Language Instinct*, "Dog Bites Man" will not make the newspapers, but "Man Bites Dog" most certainly will. Same words, different meaning. If Kanzi simply made associations between the string of words and an action, then both sentences would elicit the same response. But if he understands the syntax—in other words, he comprehends language—the two sentences would elicit different responses.

A recent study involved a statistical analysis of Kanzi's performance using the videos of Savage-Rumbaugh's many experiments with Kanzi. (Almost all the trials published in scientific papers were video recorded and are available online; see earlier.) Researchers concluded that the results cannot be explained by Kanzi's making random choices. Kanzi really seems to understand English, including grammar.

Kanzi shows that comprehension is possible in a nonhuman primate. And he is not the only nonhuman primate capable of understanding human language. Other apes that performed as well as Kanzi were similarly reared in a language-rich environment from an early age. By the time they were two and a half years old, they could comprehend forty or more spoken words. In contrast, no late starter could understand more than a few spoken words, even by the age of nine.

As human children learn language, they develop an understanding of the language they hear around them before they start to speak. They listen first, speak second. Kanzi and others like him went through the same process as human children, but of course they were never able to speak because they lack the necessary vocal morphology. In Sue Savage-Rumbaugh's mind, other ape studies failed because they assumed that speech precedes comprehension. Those studies tried to teach apes words in a language they did not understand. But expose apes from a young age to language in a way that promotes communication about topics of interest to them, and they learn just as well as human children. Granted, perhaps not quite as well, but the human–nonhuman divide certainly isn't as great as Chomsky and Wallace envisioned.

EYE POWER

Next time you are holding a baby, do the following experiment. Pull faces and see what the little one does. It's a safe bet that when you protrude your lips, the baby will protrude theirs. When you stick out your tongue, they will, too. And they will do all that without knowing what their own face looks like. Apparently as soon as sixty minutes after birth, a baby can imitate an adult's facial expression. That is how quickly babies start to build connections between themselves and other human beings. They have to, as their life literally depends on the relationship they have with others.

It is easy to fall into the trap of thinking that such a connection between an infant and the adults around it is uniquely human—inspired, in a way, by the infant's need to be cared for by many adults. You'd be in good company if you did, as for a very long time face-to-face gazing and imitation were considered by leading child psychologists to be restricted to humans. Regardless of the good company, though, you'd be wrong. It took the Japanese psychologist Tetsuro Matsuzawa twenty-two years to show that chimpanzee babies are interested in facial expressions and will imitate those expressions when given the opportunity.

Matsuzawa and Savage-Rumbaugh are kindred spirits. Instead of seeing a nonhuman ape as an object of study, Matsuzawa—as Savage-Rumbaugh did with Kanzi—treated the chimp Ai as his friend and partner. Ai arrived at Kyoto University's Primate Research Institute as a one-year-old, a victim of the now-illegal trade in chimpanzees. To connect with Ai, Matsuzawa started to behave much like a chimpanzee. Chimp and human groomed each other, hugged and hooted together. In this way, Matsuzawa gained the trust of Ai—so much so that when she gave birth to a son, Ayumu, Matsuzawa was given unprecedented access to the baby chimp. His intimate relationship with Ai and Ayumu allowed Matsuzawa to show that baby chimps, too, show an inordinate interest in faces, even if the face belongs to a human.

When another captive chimp living in the same research facility rejected her newborn daughter that then had to be raised by humans, a student of Matsuzawa, Masako Myowa, took the opportunity to test if human-raised chimpanzees are even better than chimp-raised chimpanzees at imitating human facial expressions. And indeed, the human-raised chimpanzee baby behaved almost like a human baby. When watching a human pull faces, she would pull similar faces. Yet again, it seems we humans should be pushed off our self-erected pedestal.

Well, not quite, apparently. The chimp babies' interest in imitating other faces quickly faded, while human babies only get better at it. But the key point is that our ability to bond rapidly with others at a young age is a matter of degree, not of kind. This should no longer be a surprise to you. We, too, are the product of evolution, and natural selection builds upon our ancestors' innate ability to forge bonds between infants and adults.

Science being the constantly self-correcting process it is, I need to point out that the idea young infants are already capable of imitating faces has been challenged. Follow-up studies in 2016, using a much larger number of infants, failed to show that the babies tried to mimic the facial expressions of the adult. They pull faces, that is for sure, but what kind of face they pulled was not necessarily related to the face the adult had pulled. There seems to have been a bit of wishful thinking among the researchers studying these young babies. Science remains a human enterprise, after all. It takes more time than initially claimed for young children to respond by imitation to others in their environment. What stands, though, is that young babies are intrigued by human faces.

But what has all that to do with the evolution of language? Earlier I called human babies master mind readers. They "know" how to manipulate the adults around them so they can be well cared for. While other helpless infants also need to ensure they are not abandoned before they can look after themselves, human babies take it to the extreme, laying the groundwork for the basic requirements one needs for language to be an effective communication tool.

By mimicking the facial expressions of others, a human baby learns about and builds connections between the other and the self. If this sounds a bit wishy-washy, think of it this way. Because a baby doesn't know what its own face looks like, it can gauge the relationship between the other's face and its own only by using feedback from the way its face feels when it pulls faces in response to another face. In other words, it relies on its body awareness, an example of *proprioception*. Proprioception is based on receptors found in skin, muscles, and joints, and it assists us in sensing where our limbs are relative to our trunk or how much force we need to achieve something. Thanks to your proprioception, you can find your nose even with your eyes closed. And if you've ever seen a baby whacking itself in the face while trying to stick its little fist in its mouth, you know what proprioceptive feedback is. Through trial and error, the baby improves its hand-mouth coordination until it knows where its hand is relative to its mouth.

Connecting with others is an excellent start on the road to mind reading. The next step, then, is to develop a theory of mind: the understanding that others have thoughts, beliefs, and needs similar to but different from your own thoughts, beliefs, and needs, thereby building a mental model of how the people in your environment are likely to behave. Toddlers too young to speak are surprisingly capable of inferring what another person wants. We know this because of some beautiful experiments on eighteen-month-old toddlers by the U.S. psychologist Michael Tomasello.

In one of Tomasello's extraordinary experiments, the toddler's caregiver quietly sits in the corner of a room while the little one explores the space. Then the door opens and a stranger walks in, carrying a pile of papers. His exaggerated body language makes it clear he is looking for somewhere to put his papers. Seeing a cabinet, he walks toward it. All this time, the toddler intently looks at the stranger, trying to figure out what is going on. Having arrived at the cabinet, the stranger then tries to put the papers in the cabinet without first opening the door. The toddler continues to observe this not-so-bright stranger as he keeps banging against the cabinet. Clearly, the stranger needs help, and so the toddler waddles over to the cabinet and opens the door. Having opened the door, the toddler looks up at the stranger as if asking "Is this what you want?"

In another version of the experiment, the stranger wants to hang up some clothes on a clothesline but drops the clothespin. Apparently unable to realize that the clothespin should be picked up, the stranger simply stares at it, reaching out toward the clothespin but without making any attempt to pick it up. You probably guessed the toddler's response. The toddler picked up the clothespin and handed it to the incompetent adult, again checking the stranger's face to see if indeed it is what he wanted.

The toddlers in Tomasello's experiments could not yet speak, and while the adults could, they didn't do so. Thus, all communication between toddler and adult was nonverbal. By exploring the stranger's body language and facial expressions, the toddler correctly inferred the grown-up's intentions and mental state. The toddler had no choice but to rely on nonverbal communication, but even the most verbose of us still uses silent communication. Verbal communication was simply built upon the nonverbal system we inherited from our ancestors. Before words came mutual understanding; otherwise, how would one know that the sound the other makes really refers to a particular object?

Words only make sense if there is mutual understanding between the producer of the words and the listener. And mutual understanding, in turn, relies on the ability to connect to the other. Babies begin by making connections between themselves and others around them, particularly those who care for them. As toddlers, they figure out what is going on in the minds of others, relying heavily on body language and facial expressions. All the time, they work to connect the relationship between the sounds made, the speech, and the meaning of the sounds. At the age of about six months, a toddler starts to recognize words in the language or languages they hear around them. They start to perceive the sounds specific to their native language or, if they are lucky, languages.

If you ever tried to learn a language as an adult, you will know how difficult it is initially to figure out where one spoken word ends and another starts. A native speaker produces a string of unfamiliar sounds, and you can't make head or tail of them.

Only by immersing yourself in the language will you start to recognize the specific tonal sequences associated with particular words. Most likely today, you will use an app—Duolingo or similar—to learn words and build a vocabulary. As you practice the words, you become familiar with their sounds and begin to recognize them within spoken sentences.

Babies need to rely on what they hear alone; there's no baby version of Duolingo. The more they are exposed to language, the quicker they start to recognize words. Words they hear most often are also the ones they learn the quickest. A baby's brain is an excellent statistical device. By listening, the baby learns which sounds are more likely to be paired with other sounds. If they often hear the phrase "pretty baby," their brain will figure out that the syllable *pre* is more likely to go with *ty* than with the syllable *ba*. A baby will learn a language only if it is exposed to language in a social context. Just listening to language via an audio recording, say, doesn't work. And this, I bet, is because babies rely heavily on nonverbal communication to figure out what words mean.

This brings us finally to something that *is* uniquely human. The whites of our eyes: the *sclera*. Even as adults, we humans spend a lot of time looking into the eyes of others. (Not only into the eyes of the baby we are holding.) We are so sensitive to others' eyes that simply hanging up a photo of two human eyes increases the likelihood people will pay their share into an honor box, such as for admission or home-grown produce. And if you have ever seen the *Mona Lisa*, you feel her eyes follow you around. The main explanation for why we are the only species with sclera is that the contrast between the sclera and the much darker iris makes it easier to infer in what direction someone is looking and what they are currently interested in. And this helps us understand what others are thinking or wanting. Which in turn makes humans both better communicators and cooperators—or so goes the idea.

Using both humans and chimpanzees as test subjects, a team of Japanese researchers decided to test the veracity of the eye-gazing or cooperative-eye hypothesis. Human and chimpanzee test subjects were shown photos of faces of both humans and chimpanzees that were either species-specific (in other words, human eyes with sclera or chimpanzee eyes without sclera) or modified (human eyes without sclera or chimpanzee eyes with sclera). The subjects then had to determine the direction the eyes were looking. While it took much longer for the researchers to get the chimpanzees to understand the experiment (no surprises there), once they were trained, the chimpanzees and humans were both better at figuring out in what direction the eyes were looking when the contrast between iris and the rest of the eye was starker. The sclera really do help one determine the direction of the gaze of another individual. And that means it is entirely possible that sclera evolved to enhance human nonverbal communication. Possible, but not proven.

I like the idea. In my youth, I used to be a defender on a handball team. I don't like to boast, but I was pretty good. Without looking straight at her, which would have given away my intention, I could keep track of the attacker nearing the goal from the corner of my eye. When she then tried to score a goal, thinking I had not seen her approach, I could easily intercept the ball. A slightly different version of the same trick works in sports like tennis or badminton. Gaze at where you are not aiming, and chances are your opponent will be fooled. That's because we have learned to read the eyes of people we interact with. It's also a reason why reflective sunglasses are sometimes disconcerting (and why poker players often wear dark glasses).

"The eyes are the windows to your soul..." Shakespeare once wrote. Charles Darwin was less poetic but made a similar point. Not so much about eyes but about the whole face, of which the eyes are an important component. In his 1872 book *The Expression of the Emotions of Man and Animals*, Darwin hypothesized that the facial expression we show when experiencing strong emotions—emotions like disgust, fear, happiness, or grief—has its origin in a simple sensory reaction to our environment. Our emotions are revealed by our faces because of involuntary contractions of the facial muscles in response to our mental state. Such outward signals of internal states are, Darwin suggested, left over from our evolutionary past. In our animal ancestors, such signaling served a function, staving off a competitor, threatening an opponent, or what have you, but that function is lost in humans. Our faces reveal much about our inner self, whether we like it or not. The next time you are overcome with sensory overload—say, smelling something really disgusting—try to keep a straight face. I bet you won't be able to.

If you've seen an episode of the TV series *Lie to Me*, you know about human lie detectors—people who can read the tiniest of facial expressions. The main character, Dr. Cal Lightman, is an expert in reading facial expressions and body language. He relies particularly on the interpretation of microexpressions—involuntary facial expressions that last only the briefest of time. The character of Dr. Cal Lightman is based on Dr. Paul Ekman, an American psychologist and pioneer in the study of emotions and facial expressions. Dr. Ekman now runs a training program in emotional awareness.

Taking TV shows and training programs with a grain of salt, I argue nevertheless that our facial expressions serve a function: They form an essential part of nonverbal communication. Effective communication relies on honesty and trust. Words are of no use if they do not accurately reflect the speaker's intent. By paying attention to the body language and facial expressions of the people we interact with, we form a reliable impression of their meaning. If we trust them, we are more likely to take them at their word. In fact, we so rely on trust that we have evolved self-deception—believing in a lie—as evolutionary biologist Robert Trivers argued in his book *Deceit and Self-Deception*. If you really believe in what you are claiming, chances are others do too.

LANGUAGE AS A VIRUS

Take two words: *pretty* and *pulchritudinous*. Both have the same meaning: attractive; but one is easy to remember while the other is decidedly not. Seems obvious that someone is most inclined to use the word that is easiest to remember and easiest to pronounce. If words were subject to an evolutionary process, we could say that *pretty* is much more likely to survive and reproduce than *pulchritudinous*, which would be used only by people trying to show off. If there aren't enough people wanting to show off, *pulchritudinous* will most likely go extinct. We could even take an evolutionary approach to language itself. A language that has a structure—syntax—that can easily be learned by young children (and, one could argue, by foreigners) is much more likely to take off than a difficult-to-learn language.

Perhaps linguists are looking at language the wrong way. We shouldn't be searching for a language-acquisition device à la Chomsky, but instead see language as more akin to a virus. A language that spreads easily from brain to brain is more likely to stick around. Because language depends on language learners—children—language must be tuned to the brain of children. In a way, one could say that language depends more on humans than we depend on language. If all of humanity went extinct, so would all the languages of the world. The opposite is not the case, although humanity is likely to change in the absence of language.

Sounds far-fetched? I'm not alone in thinking of language as a virus adapting to a child's brain; the idea is quite compelling. Terrence Deacon argued in his book *The Symbolic Species: The Co-Evolution of Language and the Brain* that a child's brain poses two constraints on language. The brains of young children still need to learn to make associations between sound and the thing that the sound is referring to. They also have an unreliable short-term memory. In other words, a child's brain is rather unreliable if you are an entity that depends on a brain to be spread to other brains. Just like a virus needs to overcome the immune system of its host in order to reproduce and spread, language needs to overcome the constraints of a child's brain. The better the structure of a language is attuned to the brain of a child, the more easily it will spread. A baby brain is not as good at learning language as language is good at using a baby's brain to be transmitted.

If we see language as something well adapted to the minds of our infants, we also have an alternative explanation for why it is so hard for adults to learn a new language. In a way, it seems paradoxical that the older we get, the more difficult it is to become fluent in another language. After all, our adult brain should be better at memorizing and at making associations—both necessary for learning anything new. The traditional explanation for this assumes there is a critical period during which the brain is particularly sensitive to anything linguistic. For spoken language, that period runs from about seven months to four years of age. An often-heard rule of thumb is that it is almost impossible to speak a language without some taint of your native tongue if you learned that language after the age of seven. But if languages are designed so they are easiest to learn by a young brain, it follows that an older brain has more difficulties figuring out the underlying structure of the language.

Always the brilliant naturalist, Charles Darwin made similar suggestions in *The Descent of Man*. He, too, saw similarities between "the struggle for existence" and competition among words and syntax for transmission into infant minds. He also understood how language changes, diversifying from one original language into the many thousands we have today. Just like one species slowly evolves into more than one species, a language changes and multiplies over time. Even though I consider myself fluent in English, I'm the first to admit that some Australian accents defeat me. When I had only just moved to Australia, I needed my husband to communicate with tradespeople. Before then, I lived in northern England, in the city of Sheffield, where I couldn't understand anything the locals were saying. It took me a while to train my ear to the local accent—an accent I learned to love. The rapid evolution of language occurs not because of recombination and mutations but because of linguistic innovations and errors. And at least in the case of Australia, the country's lingo has also been affected by an initial bottleneck, when the first white people arrived and the subsequent influence of languages other than English. "[T]here is in the mind of man a strong love for slight changes in all things," Darwin wrote in *The Descent of Man*. Different human populations adopt group-specific linguistic changes, and when such groups interact with one another, language diverges in a manner uncannily similar to speciation. So, we didn't need God's intervention after building the Tower of Babel; humanity is perfectly capable of creating linguistic diversity without divine intervention.

We are also good at inventing new languages—*creoles*. Creoles develop from *pidgin*, a form of verbal communication that develops between people with no common language. (Linguists do not regard pidgin as a language.) Pidgin lacks all but the most elementary grammar, and is instead made up of individual words, sounds, and gestures. A pidgin is typically a short-lived language that meets the need to communicate when different peoples meet for the first time. But when the groups remain together, the expedient pidgin quickly develops into a creole. Afrikaans is a creole language, a mixture of Dutch, Malay, Portuguese, Indonesian, and indigenous South African languages, Khoekhoe and San languages in particular. To my Dutch ear, Afrikaans is a delightful language, jam-packed with a mash of words from its constituent languages. Children are essential for the transformation, then, of creoles into full-blown languages. They are the ones who "invent" the grammar. Fascinatingly, unrelated creoles converge on a similar set of grammatical rules. These are the kind of rules that do well in the brains of young children.

LANGUAGE SCULPTS THE BRAIN

Just like a virus adapts to its host over time, a language adapts to "its" human mind through use. Maybe we should see language and human brains as being in a symbiotic relationship—a relationship in which each has something to offer the other. Language reproduces and changes through use, so we become better communicators. The better language serves as an effective means of communication, the more likely it will spread. Language and human communication both gain from the relationship. Could language and our linguistic brain have evolved in lockstep?

Our species got its large brain in part because of the repair and duplication of the gene that influences the number of neurons. But this doesn't explain the significant changes that happened to our head and face *after* our species emerged. It seems too much of a coincidence that our species' skull changed so much at exactly the same time as we found our voice. Rather, perhaps the advent of language led to changes in the very structure of our brains and skull. While we cannot transport ourselves back in time to study if, and if so how, our species' brain changed as we got better at speaking, we can study how the structure of the brain changes as children learn language.

Ceaușescu's Romanian orphans, deprived of any normal social interactions from a young age, never achieved a normal brain size, even later when they were lovingly cared for by foster parents. However, because these poor children were robbed of far more than the opportunity to develop their linguistic skills, we cannot see their plight as strong evidence of language somehow being responsible for brain growth. For that we need bilingual children.

When my daughters were in primary school, I went to a talk by a child psychologist that was about exposing one's child to more than one language. To my surprise, she argued that being surrounded by more than one language would harm the child. Therefore, she advised, parents who speak different languages should decide on one language—the language of the country they live in—and stick with it. No unfair advantage for the child of linguistically diverse parents! I remember being completely puzzled by her strong statement. In my experience, children whose parents spoke to them in more than one language simply lapped up all they heard. *Such an easy way to learn multiple languages*, I thought. My friend, who was English, and her husband, who was Dutch, fortunately took no notice of the psychologist's advice and their two children happily immersed themselves in two languages. (Later they moved to Germany, thereby adding a third language to the children's brains.)

My friend didn't have to feel guilty, as science has now shown that the psychologist was wrong. Learning more than one language makes a child's brain grow, at least some parts of it. Children who grew up learning two languages simultaneously have a larger volume of gray matter compared to those who learned only one language. Gray matter is the part of the brain where most of the information processing happens. Now, the brain is a complex structure, and some parts of the brain do show a change while others do not, but it is fair to say that language affects the structure of the brain. Which really shouldn't be a surprise. The brain changes with experience—we call that *experience-dependent neural plasticity*. My favorite example of such neural plasticity is found among London taxi drivers.

Since 1865, aspiring London cabbies have had to pass the

Knowledge of London test before they can obtain the license needed to drive the iconic black taxis. GPS navigation apps have done nothing to change the tradition. To pass the test, aspirants must memorize a whopping 25,000 streets and 100,000 landmarks within a six-mile radius of Charing Cross and the quickest routes between them. The whole process takes years, and their dedication can be seen in their brains—in the hippocampus, to be precise, which is a part of the brain that facilitates spatial memory. The longer someone has been driving a taxi, the more pronounced the changes in the brain are. Practice makes perfect, thanks to a brain that changes itself.

Terrence Deacon is convinced that language and our brain have been engaged in a process of *to*-ing and *fro*-ing during which language has changed our brain so that our brain became better at language. Our brain changes as we learn, so it makes complete sense to assume that as our species developed its linguistic skills, the brain changed. too. There is no such thing as the language center in our brain. We know this mainly from people who have suffered injuries to particular parts of the brain. The effect on their linguistic skills depends on the exact location of the damage. The different parts of the brain stay in touch via *neurons*—cells specialized in long-distance communication. As the brain develops, the neurons send out something akin to tentacles, called *axons,* and these axons form connections between the neurons in different parts of the brain. Long-distance connections allow spatially separated parts of the brain to influence each other—to such an extent that Deacon thinks the brain actively participates in its own construction.

Then there is the shape of the brain. The size of our species' brain did not change in the last 100,000 or so years, but its geometry did, leading to the typical modern human head and face, which differs significantly from those of extinct human species: think flat face and bulbous head. Scientists recently discovered that the way we think, feel, or behave is associated with activity patterns in the brain that span the entire brain and not just part of it. Instead of seeing the brain comprising different parts, each with its specific function, research now points toward a much greater integration of the brain. That means parts of the brain formerly seen as distinct now appear to be far more connected—via electrical waves traveling through the brain. If language changed our brain, I bet it changed our brain's geometry, too. Our newfound ability to make intelligible sounds reshaped the brain so that the sounds became even more intelligible. Modern *Homo sapiens* was born.

◆

We became smart enough to talk, but talking and breathing share some crucial tools. How did we become capable of doing both at the same time?

I recently attended a performance of Vivaldi's music, which included a piece he wrote for the Ospedale della Pietà, an orphanage for girls. The singers, no longer girls, illustrated a unique human trait: the ability to control their breathing to produce the most exquisite sounds. I don't want to diminish the artistry of professional singers, but in essence singing is an exaggerated form of speech, made possible because we can control the flow of air over the larynx. Of course, the original function of breathing was not to allow us to sing arias but, rather, to keep oxygen coursing through our bodies while removing carbon dioxide. Moreover, our control of our breathing is certainly not absolute. Who hasn't, as a child, competed with a friend to see who can hold their breath the longest? At some stage, no matter how much you want to win, you just have to gasp for air. As the level of carbon dioxide in your blood increases, chemical receptors in your brain and neck activate that part of the brain that controls the movement of the diaphragm. Whether you want to or not, you will take a deep breath. Yet, we can learn to increase the length of time we don't breathe, irrespective of what the brain says. The current record is held by Branko Petrovic, who held his breath for 11 minutes and 54 seconds. Petrovic is part of a growing group of "static apneans," or breath-holders. But more ordinary people can train themselves to hold their breath for far longer than is natural. Actors can learn to do it if required for a particular scene.

Many other mammals can stay underwater for extended periods and hold their breath for much longer than 11 minutes and 54 seconds. But the physiology of aquatic mammals is different from ours. They are adapted to be able to dive for extended periods. Not so for us land critters. Yet, our species managed to take control of an essential physiological act—breathing—and mold it to a new purpose. That purpose is to speak. And to sing, but that was most likely secondary.

Here is an interesting twist. The parts of the brain that allow us to control our breathing—the corticospinal and corticobulbar tracts—also control our facial muscles. These are muscles we need to control for our speech to make sense. When we are asleep, another part of our brain—the medulla oblongata—takes control of our breathing. Two bits of brain with complementary functions. It seems we coopted a part of our brain to improve our control over the sounds we make.

A LITTLE (ADAPTIVE) LEG UP

Our ability to evolve superb communication skills did not come out of the blue. (*We* didn't come out of the blue.) Before us, there was a long line of ancestors, each with a set of traits that gave natural selection something to play with. Our ability to speak was the result of a short series of genetic and anatomic flukes that set the stage for runaway selection.

Sometimes our evolutionary history reminds me of one of those stair-climbing contraptions you find in a gym, where the only way to prevent serious injury (trust me, I know) is to just keep putting one foot in front of the other. Neither Wallace nor Chomsky could see how a little bit of language would give us a leg up in evolution. But I have argued that our childcare problem was incentive enough to keep climbing. Our tiny babies are in need of extraordinary levels of help. They compete with siblings and others for attention and for resources. Being slightly better at vocalizing your needs than the others may just be enough to put you ahead of your competitors. Mothers, too, would benefit from being just that little bit better at enticing others to help with raising the children. With every improved fragment of speech, and every moment of shared understanding, we raised the likelihood our babies—and our species—would survive. Yes, human language makes us fundamentally different from other apes and animals. But all the innovations required for abstract, compositional language can be explained by natural selection. What makes us special simultaneously confirms that we're part of the natural order.

seven

OTHER MINDS

Not being too fond of cities, I've spent the past twenty years or so commuting by train from a small town on the central coast of New South Wales, Australia, to my laboratory and office at the University of Sydney. The hour-plus ride has always given me plenty of time to observe the diverse behaviors of my fellow humans. Some passengers immediately fall asleep in their favorite seats. Others listen to music, or perhaps a podcast, as they watch one of the most picturesque landscapes in the world amble by. I still smirk when I think about the woman and man who flirted shamelessly every morning on the platform, going their separate ways as soon as they left the train, only to pick up the next day just where they left off. For all our differences, this temporary community always has one big thing in common: our species. You might think that goes without saying, but the truth is it could have turned out so differently. Allow me to imagine an alternative train ride on the 7:23 a.m. from Woy Woy to Sydney Central.

The passenger sitting across from me looks so small, but then again, his diminutive head is in perfect proportion to his body: a fine specimen of *Homo floresiensis*, also known as The Hobbit. And that flirting couple? I can understand the mutual attraction, since they're both of rather robust build and similarly hirsute. Their smiles showcase magnificent, powerful teeth. She's a clear example of *Homo neanderthalensis*, whereas I suspect he belongs to *Homo denisova*, given the reddish tint of his hair. There in the corner, I spot a young *Homo erectus*, holding a finely shaped stone tool in her hand and keeping to herself. Next to her sits a *Homo heidelbergensis* boy, recognizable by his large head, softly humming to himself, while trying to ignore the girl beside him. Sounds like science fiction, sure. But all these species were around when we evolved. Some even mated with us, leaving traces of themselves in our genomes. Why, then, are we the sole survivors?

To my mind, the answer as to why our species displaced all other humans can only be language and the nuanced collaboration that it makes possible. But not everyone agrees.

THEN THERE WAS LANGUAGE

In *How Language Began*, author and linguist Daniel Everett makes the controversial claim that our ancestors had been chatting with each other well before our species evolved. In which case, language couldn't have been the distinguishing factor in our species' triumph over all others. Everett's proffered evidence is that a possible offshoot of *Homo erectus*—dating back roughly 1.89 million years—got stranded on the Indonesian island of Flores, where it evolved into The Hobbit, *Homo floresiensis*. To get to Flores, *Homo erectus* had to cross the Wallace line, the deep-water channel that prevents the movement of most animals (and to a lesser extent plants) between Asia and Australasia, resulting in strikingly different faunas between the two regions. The deep-water channel runs down through the Lombok Strait that separates Bali and Lombok. Flores is east of Lombok and thus east of the Wallace line, so that the ancestor to The Hobbit had to cross the deepwater channel. And for that, Everett argues, our early sister species must have used some form of language to plan the trip and build their rafts.

Others have argued that a massive storm could have blown the ancestors of The Hobbit to Flores from adjacent islands, clinging to debris. But my social insects illustrate a deeper problem with the early language idea. As we've seen, honeybees can easily coordinate their move to a new nest site through dance—a kind of communication, to be sure, but nothing nearly as sophisticated as our abstract, compositional language. In fact, fire ants build rafts using their own bodies when their underground nests flood from rainfall—without using any means of communication. Rafts that, of course, are disanalogous to human rafts in countless ways, but still show how little language or planning is strictly necessary for engineering projects in the wild. And what are we to think of the other large animals that got stranded on Flores? Because The Hobbit was certainly not alone. Massive Komodo dragons and half-ton elephant-like creatures are salient examples of other animals "stuck" on the island. No one is arguing they could speak or build rafts, yet somehow they, too, crossed the Wallace line. (It may be that the Komodo dragon crossed from east to west instead of from west to east. But cross the channel it did.) Only when talking about humans and our ancestors do we feel compelled to assume extraordinary circumstances. As Carl Sagan once said, extraordinary claims require extraordinary evidence. I see no such extraordinary evidence that points toward a speaking ancestor to The Hobbit.

Now might be the time to define *language* a bit more precisely. Language is more than communication, as everything alive communicates in one way or another: to organize mating or to get together for other reasons. Even bacteria communicate; they send out chemical signals that other bacteria respond to, allowing them to form large aggregations. Biofilms are a good example of such aggregations; and a good example of a biofilm is the plaque that forms on your teeth.

Language is a computational cognitive system that allows the combination of a finite number of symbols—words—to be combined, following a set of rules—syntax—to construct an infinite number of sentences. That rather technical definition translates to a simple concept, nonetheless. Language allows us to talk about everything we want to talk about, even things that aren't real, aren't present, existed in the past, or may exist in the future. According to the historian Yuval Noah Harari, that ability—talking about unreal things—is key to our success. By telling stories, our species became the unrivaled ruler of the earth, with devastating consequences. Harari's *Sapiens: A Brief History of Humankind* explains how we, in the blink of an eye, not only got rid of all other species of *Homo* but also invented, and believed in, fictional things such as nations, religion, arts, and money. Sometime in our recent evolutionary past, the cognitive revolution took place, which transformed our species from being just another *Homo* among several to one of the most powerful living forces the earth had ever seen. (Not *the* most powerful force on earth; arguably, photosynthesis deserves that honor because without photosynthesis no animal life would have been possible.) Language allowed our species to build strong social relationships through shared stories, forming large and powerful groups that were no match for the smaller groups of sister species.

Through stories, our species collectively believed in things, and that collective belief led unrelated individuals, even strangers, to interact positively with each other. Just like modern children invent passwords to limit membership in the club, our ancestors used shared stories to determine who is "in" and who is "out." Once we were able to tell stories, and invent an imaginary world, our species' behavior changed rapidly, and innovations outside of genetic changes were quickly transmitted to the next generation. In other words, cultural evolution took over from genetic evolution, swiftly transforming our species and so the world. But before we started to tell stories, we needed to invent language.

What Yuval Harari leaves unexplained is how we acquired language during this cognitive revolution. How did a little bit of storytelling lead to a boost in reproductive success? Would a half-baked story really have given our species a big enough advantage? Yes, once we had language, things would never be the same; but storytelling cannot have been the key that unlocked our species' potential.

Perhaps Harari's cognitive revolution went something like what was envisaged by Harvard professor of human evolutionary biology Joseph Henrich. In the beginning, our ancestors were already excellent learners and inventors. When watching someone use a tool, a procedure, or trick to solve a novel problem, the observer could remember that particularly awesome solution and perhaps combine that trick with some other innovation they'd learned from another person. By combining the innovations from several individuals, our ancestors enhanced their individual and collective toolkits. They became smarter. Henrich refers to this period of learning and copying as cumulative cultural evolution. Possibly at the dawn of *Homo*, our ancestors crossed a point at which cultural evolution became *the* driving force, propelling genetic evolution toward what would ultimately become us.

The cognitive revolution, then, was the result of our ancestors' increased skills and knowledge, leading to the need for the brain to catch up to be able to hold so much information. That cognitive revolution was driven by genetic changes, by selecting for individuals who had the right complement of genes for higher cognition. This relentless selection for higher cognition led to an increase in brain size, until the brain became so large it started to hinder babies' births. Around 200,000 years ago, the increase in brain size came to an end. *Homo sapiens* had arrived.

Henrich has used the tragic fate of many explorers of new territories to illustrate the importance of culture. The European explorers of the eighteenth and nineteenth centuries had an expansive cognitive toolkit, knew how to use tools, and had language. Yet, many succumbed in unknown lands through a lack of food or water, while surrounded by both. Why? Because they did not have the cultural knowledge to know what to eat, how to prepare it, and where to find water. Key to our species' success, Henrich argues, was the need to acquire, store, organize, and re-transmit the growing body of information created by the cultural evolution that allowed us to adapt to our environment, outside of genetic change. And all that cultural information is stored in a large number of brains—the brain collective. Somehow, somewhere during that cultural revolution, language appeared as another means, a very effective means, of transmitting cultural knowledge.

The point that Henrich makes about the need to be taught what to eat and where to find it is an interesting one. But that need is not restricted to humans. The cassowaries that regularly visit our property (or rather, visit a place they have long inhabited and where we have recently come to reside) keep track of what tropical trees are producing fruit and when. Almost like clockwork, the magnificent birds appear for as long as the tree produces its fruit, or until something better somewhere else is yielding its fruit. Dad looks after the chicks, and in so doing, teaches the young ones the idiosyncrasies of their environment, saying something like "In November we go to the creeks for the river figs. When they run out, we search for the blue quandongs that grow on the higher ground." The chicks spend about a year with their Dad, which is enough time to learn the annual cycles of the fruit trees in their environment. If adult birds are relocated to another area, as they sometimes need to be, they may struggle to find food for a year or two.

The authors of *The First Idea*, Stanley Greenspan, M.D., and philosopher psychologist Stuart Shanker, go so far as to say that language and symbolic thought cannot be explained at all by genetics or natural selection. Instead, those capacities are socially reinvented with every successive generation. In the space of a childhood, all our adaptive skills, accumulated collectively, over tens of thousands of years, are re-learned. In a process reminiscent of Ernst Haeckel's (incorrect) idea of ontogeny recapitulating phylogeny—the thought that as an embryo develops it retraces its evolutionary history—Greenspan and Shanker see our own evolutionary history reflected in each child, as it develops from a tiny mind reader into an accomplished storyteller. Our cognitive abilities developed not so much as a revolution but, rather, as a gradual process, through social interactions passed on through the generations via learning. And again, somewhere, at some time during that process, language made its appearance.

Regarding these arguments, I agree that cultural evolution has changed us more rapidly than natural selection ever could have. But these careful scientific and historical minds all put the cart before the horse. Cultural evolution took off only once we had the ability to speak, not vice versa.

THE APE THAT GOSSIPED

Here's another explanation for our exceptionality: gossip and status.

The zoologist and evolutionary biologist Richard D. Alexander spent his lengthy career trying to explain why animals behave the way they do. And that includes humans. Being a biologist, Alexander understood what the real problem is that we need to explain: cooperation among non-family members. Something we humans are exceptionally good at. But why is cooperation with strangers something that needs explaining? After all, social insects do it all the time, as well as many other animals, but they mainly interact with family. And because your relatives share a lot of your genes, cooperating with relatives is a way to help get your genes into the next generation. As the British evolutionary geneticist John Burdon Sanderson Haldane allegedly once quipped, "I would lay down my life for two brothers and eight cousins." The closer the relationship, in genetic terms, the higher the chance two individuals share a large proportion of their genes. Helping unrelated individuals means you are spending time and effort without any of your genes being passed on. And that is something natural selection does not approve of.

That last sentence is, of course, a bit silly. Natural selection does not "approve" of anything, but the point is that there is no straightforward mechanism by which natural selection can select for a behavior that has no effect on the way genes are transmitted to the next generation. That is why Alexander wanted to find a not-so-straightforward explanation for why we are so good at cooperating with others. In doing so, he built on an argument that Darwin had made in *The Descent of Man,* where he wrote "A tribe including many members who, from possessing in a high degree the spirit of patriotism, fidelity, obedience, courage, and sympathy, were always ready to aid one another, and to sacrifice themselves for the common good, would be victorious over most other tribes; and this would be natural selection." Strength in numbers. But how to get there?

The first step is to stop being aggressive toward members of your group. Caring for helpless infants and having access to plenty of food had a lot to do with our tendency to be nice to others, at least others who are part of "our group." Some don't think that was enough to have turned us into peace-loving do-gooders. Anthropologist Richard Wrangham, in *The Goodness Paradox,* argues that it is violence that caused our peaceful nature. To maintain social cohesion, we became excellent at actively getting rid of bullies, by executing those who refused to conform. Not being a good cooperator? Off with your head. The elimination of aggressive individuals within the group, through capital punishment, has led to peaceful coexistence within large groups of unrelated individuals. But at the same time, the need to be executioners gave us the capacity for ultra-violence. Hence the paradox.

What if, instead of killing nonconformists we talked each other into being nice?

That is the idea behind Richard Alexander's reputation building—ensuring that others see you the way you want to be seen so you receive all the help you require. The more favorable someone's opinion of you, the more likely you are to receive benefits such as care and resources. Or, the alternative, if you are seen as a nasty piece of work, you are less likely to be helped or fed by your group. There are good examples of the selective allocation of resources depending on an individual's social status within the group. Known as a kind, empathetic person, always willing to assist others? Check. Excellent gatherer of the most difficult foods? Check. Repository of valuable information? Check. Lazy? Thumbs down. You get the gist.

Your reputation not only precedes you it can also be dialed up or down by gossip. The more people praise your virtues, the better your reputation and the more people would want to cooperate with you. In this way, gossip helped our ancestors foment social cohesion within a group, turning them into super cooperators.

Cleaner fish also care about their reputation. These fish do exactly what their name suggests—most of the time at least. They clean other fish, removing and eating ectoparasites and dead skin from the so-called client fish—fish that specifically go to the cleaner fish's hangout to be cleaned. To make it easier for client fish to find a cleaner, cleaner fish hang around the cleaning stations. There, the cleaners make it known to their potential clients that they are available for cleaning by performing something akin to a dance. A dancing cleaning fish.

As the saying goes, the client is king, and so it is the client fish that "selects" a cleaner. What is it that a client fish looks for when trying to find a good and reliable cleaner? One factor is that the cleaner fish is not tempted to bite a chunk out of the client's body. There is always the temptation, as a cleaner fish, to quickly take a nibble out of your client. Ectoparasites and dead skin are not as tasty as an actual bit of fish. Somehow, the cleaner fish has to signal it is a reliable fish, because otherwise it won't get any clients and so no food.

In the field, client fish keep a keen eye out for the interactions between a fish receiving a clean and its cleaner fish. If the interaction ends without drama, the client fish is more likely to allow that particular cleaner fish to provide services. Reputation matters. Seen as a reliable cleaning fish? Then that fish is more likely to be chosen. That's a nice story, but to really figure out if the potential client fish indeed cares about reputation, we need to do some experimenting.

How to manipulate a cleaner fish's reputation? Take an aquarium and divide it into three parts. On both ends, place a cleaner fish. In the middle of the aquarium, put a client fish. The client fish can look at the two cleaner fish via a one-way mirror so the cleaner fish does not know that it is being observed. Now, give both cleaner fish a fake client fish. One toy fish has prawn paste smeared on it, the other has nothing so there is no reason for that cleaner fish to get near the fake fish. Cleaner fish like prawn paste, and so the lucky fish will quickly start to eat the paste off the fake fish. That looks just like cleaning. The other cleaner fish has nothing better to do than aimlessly swim around. Little does it know, though, the prospective client fish thinks that the prawn-eating cleaner fish is doing its duty, whereas the other is ignoring a potential client. If, the experimenters hypothesize, the client fish takes into account the prior behavior of the cleaner fish, it would more likely select the fish that seems to be doing what it is supposed to do.

Lo and behold, that is exactly what the client fish does. It spends more time at the end of the aquarium near the "cooperative" cleaner fish than at the end with the innocent "cheater" fish. Clients care about reputation, and they prefer a cleaning cleaner fish over one that seems to be shirking its responsibilities.

Both the cleaner fish and the client fish benefit from the relationship, although the cleaner fish could get more if it behaved differently. If it bites its client, it may get some extra nutrients but that spoils future relationships. Short-term gain, long-term pain. In any relationship, cheating is always an option; that is exactly why it is important to keep track of your potential partner's past behavior. And why it often pays to be honest.

Time to return to Richard Alexander's question: What turned us into super cooperators?

Our kinder closest living relative, the bonobo, can perhaps provide insights into how our pre-verbal ancestors might have behaved toward nonfamily members. We often forget it is not just the more violent common chimpanzee that is closely related to us; our genetic distance from chimps and bonobos is equal, but because bonobos live in one of the most dangerous and underdeveloped countries in the world, they are much harder to study logistically than are chimps. I also suspect that some researchers feel greater affinity with the explosive disposition of the common chimpanzee than they do with the more mellow bonobo. It's more a "man the hunter" than a "make love, not war" narrative—an evolutionary history more driven by male aggression, and the means to curtail it, than by peaceful cooperation. Most evolutionary biologists, in fact, are more comfortable seeing evolution as rife with conflict. But survival of the fittest does not necessarily mean survival of the strongest.

Bonobos regularly share food, groom, and form coalitions with members of other groups. Here, too, reputation matters. The more cooperative a prospective partner is, the more likely that partner will be chosen. Bonobo behavior shows us that cooperation with unrelated individuals can evolve without gossip, but gossip—or language in general—most certainly allowed human groups to swell and become super cooperators. Remember Darwin's words: Such large groups, through language, would be victorious over other groups. But in this instance, those other groups were not members of our own species; they were other species of *Homo*.

WHEN *WAS* THERE LANGUAGE?

The anatomical bits that make language possible do not fossilize, so it is not easy to say when it began. The best we can do is to come up with educated guesses. Many authors have made the not unreasonable assumption that the rapid change in human behavior around 100,000 years ago is circumstantial evidence for the appearance of at least some form of language. That is when we start to find more sophisticated tools, signs of long-distance trading, and the appearance of symbolic artifacts, such as art and jewelry.

Art and jewelry are useless objects, made because they are considered beautiful or make the wearer more beautiful. They can also represent something that is revered. As they say, beauty is in the eyes of the beholder, but art and jewelry are in the mind of the maker and the observer. They are symbols. To come up with symbols requires symbolic thinking. The sentences written on this paper comprise symbols—we call them words. If words are symbols, we needed to invent symbolic thinking before language was possible. The presence of artifacts that can be interpreted as art or jewelry are therefore evidence of symbolic thought, and, by extension, are indicators of some form of language.

Sometime between 65,000 and 160,000 years ago, our ancestors started to make more sophisticated tools—tools that were so complicated it seems unlikely an apprentice could rely on observation and imitation to learn the trade. The only way to effectively learn how to make these complex tools must have been through explicit instruction, and for that, language was needed. I challenge anyone to give explicit instructions without language.

Further evidence of the emergence of human trade lies in the details of goods. Some of the raw materials used for decorations or tools were not found in the same location as were the artifacts. This implies that either the artifact itself or the material from which it was made had been acquired from far afield. These goods or materials were traded. And for trade, one needs trust in strangers and an effective means of communication—again, language.

The timing of our species' invention of new tools, trade, and art coincides with the massive changes to our skull. These were the kind of changes that gave us our modern *Homo sapiens*-specific head and, through anatomical changes, made language possible. Coincidence? I, and many evolutionary anthropologists in particular (who combine biological and cultural evolution) don't think so. But we need more than circumstantial evidence if we want to convince the skeptics. Luckily for us, there is a way to look more directly at the likely origin of language.

Just as evolutionary biologists can estimate the time since two species last shared a common ancestor by looking at the accumulation of mutational differences between their two genomes, so too can we estimate the divergence among languages by the changes to the diversity of *phonemes*. Phonemes are the smallest units of speech that distinguish one word from another. In English, the phoneme /k/ is found in words like *cat, scat, kit, skit*. The diversity in phonemes then gives an idea of the number of distinct units of sound one can decipher—vowels, consonants, and tones—in a language. That diversity is language specific. Modern African languages are the most phonemically diverse. The smallest phonemic language diversity is found in South America and Oceania—some of the most recent areas to be colonized by our species. !Xun, a language spoken in southern Africa, has 141 phonemes. Rotokas, a language spoken in New Guinea, and Pirahã, a South American language, both have only 11 phonemes.

Think of a single phoneme as the equivalent of a codon in DNA (a codon is the three "letters" that code for a protein). The more different codons, the larger the diversity of proteins that can be produced. Similarly, the higher the number of phonemes a language has, the more that can be said. Or at least, one can express things in a more nuanced way.

Imagine what happened as some of our fellow *Homo sapiens* moved out of Africa. These founders took with them a subset of all the genes present in the African population, so to this day, all non-African human populations have smaller genetic diversity relative to African populations. (A reduction in diversity caused by a small number of individuals settling into a new area is known as a *founder effect*.) As the immigrants started to spread around the world, their phonetic diversity declined, through the same founder effects. Each time a new area was colonized by a small group of humans, some phonemes were lost. At least that is the reasonable assumption made in a study that uses phoneme diversity to estimate when language evolved. New phonemes also appeared as populations became isolated from other groups of humans.

Phonemes evolve slowly, just as mutations accumulate slowly, but unlike mutations, phonemes transmit culturally so their gains and losses are unlikely to be caused by changes in cognition of the human populations under study. The rate at which they are lost, and the new ones appear, make phonemes the perfect vehicle for tracing the timing of the first words. From that, it follows that the older the language, the higher the diversity in phonemes. And then the "only" thing you need to do, by backtracking, is to figure out how long it probably took to get to the phonetic diversity we find in Africa today, taking into account the rate at which new phonemes appear.

Using that logic, we can say that language is old, going back to the Middle Stone Age, 50,000 to 280,000 years ago. But it's not so old as to turn *Homo erectus* into a chatty relative.

How chatty were our closer relatives, the Neanderthals and Denisovans? I hinted at the answer in Chapter Six, in my discussion of the "human" version of the *FOXP2* gene. Neanderthals had the same version of *FOXP2* as we do. But that doesn't mean they could speak like we do, because we've learned that *FOXP2*, while important for sound production, is not a gene "for" language, let alone *the* language gene. Again, it is unfortunate that the parts of our anatomy that directly enable speech are too soft to fossilize, so we cannot look at the remains of our fellow *Homo* species to say with certainty if they could talk. Luckily, there are other clues.

The length of the neck and the size of the vocal cavity are two related cues as to whether a *Homo* species could talk. Neck length and vocal cavity size are related because they both affect structures and cavities that form the vocal tract—the larynx and the pharynx, and the nasal and oral cavities. By controlling the position of the tongue, lips, and larynx, we can change the shape of the vocal tract, affecting the sound that comes out of our mouth. To be clear, I do not argue that Neanderthals and Denisovans were incapable of making vocalizations that could improve communication over that of, say, chimpanzees. My claim here is that they lacked the *precise* control necessary for phonemes, the functional units of language. One cannot distinguish between the phoneme /t/ in *tea* and *the* without control of the position of the lips at the beginning of each word. (You can actually feel the difference just thinking the two words.) Other phonemes rely on control of other parts of the vocal tract. Being a native Dutch speaker, I am good at making the harsh /g/ phoneme, as in saying the name of the famous Gouda cheese, which I do by closing my vocal cavity with the base of my tongue and pressing air through a smaller space. My husband, who is a native English speaker, is unable to properly pronounce the hard /g/ we Dutch speakers (and speakers of Arabic, Hebrew, and German) are so fond of. It's not because there is something wrong with his speech, but because English has no need for a super-harsh /g/. For the soft /g/, as in *giraffe*, I move the tip of my tongue to the top of the palate, just behind the teeth. It is quite fun to think about the ways we create subtle differences in our speech.

Precise speech also needs control of parts of our body that regulate airflow over the vocal cords. Remember the superb control of breathing by the choral singers I so admired. The singers were expert at controlling the movement of their thoracic muscles, including the diaphragm, to affect the tempo and strength of airflow from the lungs. We can all do it, regulating the tempo and strength of airflow over our vocal cords, but sadly most of us are not as accomplished as professional singers. If you want to figure out how the vibrations caused by the airflow over your vocal cords affect your speech, inhale helium. Helium is less dense than air and causes the vocal cords to vibrate at a high frequency, giving you a high, squeaky voice. A great party trick.

Who knows what level of control our relatives had over their breathing, but we can say that their neck and face do not seem to be the right shape to have the vocal-tract architecture necessary to achieve the precise control needed for exact speech. Their faces protruded too far, and their necks were too short. To obtain the ideal relationship between the depth of the oral cavity and the depth of the larynx, given their protruding face, highly verbal Neanderthals would have had their larynx sitting so low in their throat that it would have made eating difficult. (As mentioned, we cannot say much about the faces of Denisovans, as we don't have any fossils of their skulls.) As we swallow, a small U-shaped bone called the *hyoid* pushes the larynx into a position that makes it less likely for food to go down the windpipe instead of the esophagus. For that to work, both the larynx and the hyoid need to sit in the middle of the neck. If both were positioned too low in the neck, because the neck is too short, the sternum (breastbone) and collarbones would interfere with the movement of both the hyoid and the tongue, making swallowing more difficult. Hence, Neanderthal anatomy suggests their larynx could not have descended enough to give them the control of their tongue needed to allow precise speech.

Neck and face anatomy are not the only evidence that archaic humans could not speak. There are also some interesting hints to be found in their DNA. Or, to be more particular, *on* their DNA.

Most genes most of the time need to be turned off, not on. You don't want the genes that help synthesize bile being expressed in your brain or your big toe. One way in which genes are silenced is with epigenetic marks attached to the DNA of the gene to be silenced. Epigenetic marks cause contraction of the chromosome and the DNA within it, thereby making the gene invisible. It's invisible in the sense that it cannot be translated, or read, to produce protein or other gene products.

Epigenetics is essential in building a multicellular organism, organizing which genes are "on" and which are "off," depending on where in the organism they are located. Without epigenetics we could not have a brain cell or a liver cell. Both have the same DNA sequences but they differ in their epigenetic marks, so a brain cell has genes expressed that are needed to build a brain, and liver cells have liver genes turned on. Epigenetics is one reason why organisms can be genetically similar yet look very different. Their epigenetic marks change their phenotype. If one could find out which genes were marked for epigenetic silencing in human fossils, and compare this gene list to the silenced genes of modern humans, one could find out what sort of genes were likely to have had opposite on/off settings. It turns out we can do this for at least one kind of epigenetic mark: methylation.

To methylate a gene, a methyl group, comprising a central carbon atom bonded to three hydrogen atoms, is attached to a cytosine, one of the four bases that constitute DNA. The methyl group serves as a stop signal, preventing the translation of the gene into a functional product. In all mammals, but also in plants, methylation of so-called promotor regions regulates whole gene networks. Methylation determines if the row of dominoes falls (network "on") or remains standing (network "off").

Like everything else, DNA degrades over time. Sequencing ancient DNA is particularly difficult, not just because it is old but also because it is often kept under less than ideal conditions. We don't tend to find fossils in minus 110° F (minus 80°C) lab freezers, which would be ideal, but in much warmer caves. Not all is lost, though, especially if one is interested in using methylation patterns to figure out which gene networks were likely to have been "on" in your individual of interest. Here is the thing: Cytosines that were methylated degrade over time into thymines, while those that were not methylated decay to uracils. That means one can figure out which DNA sequences were methylated and which were not, even when the DNA is old. Provided the DNA is not too old, of course.

Our immediate ancestor separated from Neanderthals and

Denisovans 550,000 to 765,000 years ago, and the Neanderthals and Denisovans split from one another between 445,000 and 473,000 years ago. Seems like a long time, but we are close relatives. After all, we mated with them. Or, at least some of us did. The secret to our differences is, therefore, unlikely to be found in our genome, especially given we don't differ that much, genetically, from chimpanzees. Most of the physical and behavioral differences between us and the archaic humans are to be found in our epigenomes. While much of the epigenome is related to the state of the histone proteins that hold DNA, and is therefore never fossilized, the methylome is partially preserved.

Comparing methylomes to figure out the key differences between us and our close relatives is exactly what a team of researchers did. Which regions of the DNA from present-day humans and fossils of anatomically modern humans showed methylation patterns that are different from those of all Neanderthals and Denisovans, and chimpanzees? The results are telling.

The team, led by David Gokhman, in 2020 reported that our methylome contains 873 regions that have methylation patterns not found in our *Homo* relatives or chimpanzees. Many of these regions have no known function, but 588 were associated with known genes. Chimpanzees, too, have those genes, but their methylation pattern prevents them from being turned on. Of the 588 genes found, 56 are involved in the morphology of the face and lips and 32 are involved in the larynx. The same genes in Neanderthals and Denisovans did not differ in their methylation pattern from that of chimpanzees. I think that nails it. We have unique gene regulatory networks that affect the anatomical structures needed for the production of precise sounds.

We may have made babies with Neanderthals and Denisovans, but I don't think we had much to talk about.

A BEAUTIFUL MIND VERSUS THE CARING MIND

One doesn't need language to be able to move out of Africa. *Homo erectus* left Africa way before us, colonizing large parts of the world and leaving many different offshoot species in its wake. Our species' mark on the world was different. When we left Africa less than 100,000 years ago, it was the beginning of our unstoppable march to colonize every corner of the earth. Instead of giving rise to more species, as *H. erectus* did, we exterminated our relatives. To be fair, our species is probably too young to have given rise to new species, but our young age makes the rapid demise of our relatives even more puzzling. It is language that made us do it. Once we mastered words, we built powerful relationships that allowed us to collectively care for helpless infants, setting the scene for the many characteristics we now consider uniquely human. We do not need to evoke some sort of special language faculty to understand why we came to speak. We just need to understand why speaking gave our species such an advantage. The rest, as they say, is history.

One mystery remains. Why did brain size in our *Homo* relatives increase slowly but steadily? If I am right, language was not the driving force behind the brain's gradual increase. Neither was tool making, as the tools that earlier species of *Homo* constructed didn't change much until our species made its appearance. According to the evolutionary psychologist Geoffrey Miller, the answer is sex. Or more precisely, sexual selection and mate choice. In *The Mating Mind*, Miller argues that our brain is equivalent to the male peacock's tail, an ornament whose primary function is to woo the opposite sex. The wittiest, funniest, smartest males had a higher chance of being chosen by females, thereby spreading the genes for wit, humor, and smartness. And because being witty, funny, and smart requires a larger brain, the brains of early *Homo* species started to grow, culminating in our outsized brain. The larger the brain became, the more extravagant the means by which males attempted to entice females to choose them as the father of their children. And so seemingly useless and uniquely human traits evolved—traits such as music, art, and poetry.

Miller's argument may sound outlandish, but consider the satin bowerbird. Bowerbirds, of which there are twenty-seven species, are found only in Australia and New Guinea and are renowned for their amazing courtship behavior. To seduce a female, a bowerbird male builds a special structure, a bower, in which he displays for the female. I once stayed in a holiday cottage in the temperate rainforest of New South Wales, in Australia, when I heard strange sounds I could not identify. Squeaky, squelching, whistling sounds interspaced with something that sounded like falling bombs. When I went outside, it wasn't hard to find the culprit. A male satin bowerbird was practicing his song and dance in isolation, as there was no bower or a female in sight.

Once confident about his dancing and singing ability, the male would start constructing his bower: two parallel walls of sticks stuck in the ground, adorned with anything blue the bird can find. These days that includes bottle caps, drinking straws, clothespins, and what have you. Nothing blue is safe from a satin bowerbird male come mating season. The famous O'Reilly's

Rainforest Retreat, in Lamington National Park near Brisbane, in Queensland, had to give up using their blue drinking straws, as the birds would steal them all from the restaurant, even from people's drinks.

The birds' love for blue comes with a dislike of other colors, particularly the color red. And that allowed a group of biologists to do a neat experiment.

Female bowerbirds select their mating partner based on the quality of his song and dance, and the beauty of his bower. Males don't contribute anything to the young other than their sperm, so a female cannot be selecting a male based on some sort of indication of his paternal abilities. In another species of bowerbird, the great bowerbird, males construct a court of lightly colored objects—stones, bones, bits of weathered wood, shells—at one end of their bower. The objects are arranged according to size, with smaller objects closer to the bower and larger ones farther away. A female approaching the court from within the bower will be treated to an even mosaic of visual angles that creates a forced perspective illusion. The perspective also makes the male seem much bigger as he is standing at the other end of the female, in his "courtyard." The better the visual illusion, the higher the chance the female will mate with the male. It seems as if the female uses the quality of the display as a means to suss out the male's cognitive abilities. This is where the satin bowerbird male's dislike of red comes in.

The research team was keen to test the idea that the bowers of bowerbird males and their displays are a means for females to select the male with the best cognitive skills. In doing so, females would increase the probability that their offspring are better at learning and problem solving, assuming that cognitive skills have some genetic basis. The problem the researchers gave the satin bowerbird male to solve was to get rid of red objects.

When the male was out foraging, the researchers would place three red objects near the bird's bower, covered by a clear plastic container. Upon his return, the enraged male, provided he was clever enough, would lift the container and carry the offensive objects as far from his bower as he deemed necessary. The speed with which he got rid of the red objects was a measure of his problem-solving abilities. If he passed this test, the next test was more difficult. This time three objects—colored squares—would be nailed into the ground, again close to the bower. One square was blue (good color), one green (not as good as blue, but still acceptable), and one was offensively red. This time, the male could not pick up the red object and fly away with it. His only option was to cover the thing so he no longer had to look at the disagreeable color. How quickly and how well he managed to make the red square invisible was then a measure of his problem-solving ability.

Because bowerbirds live for many years (the record in the wild stands at twenty-five and a half years), the research team could match an individual male's problem-solving rank with the number of times he was selected by a female as her mate. Guess what? The males best able to quickly solve their problem with red objects had the greatest success with the females.

We can't ask a bowerbird female what it is she is looking for in a mating partner, but the experiment lends some credence to Geoffrey Miller's idea of the mating mind.

Sexual selection was Darwin's other great theory, laid out in *The Descent of Man, and Selection in Relation to Sex*. Charles Darwin was puzzled by the many traits he saw that, instead of providing its bearer with a survival advantage, seemed to hinder the individual. Most such traits appeared to be restricted to the males of the species. Sex-specific traits, or sexual dimorphism, can be so pronounced it may seem males and females belong to different species. Carl Linnaeus himself was fooled. In 1758, he described two species of duck, *Anas platyrhynchos* and *Anas boschas*, not realizing they both were mallards. The former was a female, the latter a male.

Males in sexually-reproducing species may compete for mates in two main ways. The first is directly, mostly via fighting, with the winner taking it all. Such direct male-male competition we find most often in species in which one male dominates many females. Or, second, as the bowerbird example illustrates, males compete for the female's attention. "Look at me, look at my bower! See how clever I am!" We met another such male earlier: the Japanese pufferfish and the beautiful geometric patterns he constructs on the ocean floor. My favorite by far is the tiny peacock spider, a species of jumping spider. It is only about 0.2 inches long, the size of a grain of rice, but the male's behavior makes you melt. When he meets a female, the male lifts his abdomen to reveal the most beautiful color pattern. Standing in front of the female, he starts to dance, rhythmically moving his colorful behind while waving his front legs. I highly recommend looking at a video of the male's impressive display.

With the peacock spider, we are in the second, nonviolent way of attracting a mate. At least that is Miller's argument. By selecting males who are creative, resourceful, and witty, our female ancestors' preference led to an increase in brain size. And as the brain size of males increased, so did the brain size of females, because males and females share almost all their genetic material. While I am attracted to Miller's argument, I see an obvious weakness. In his attempt to explain why we spend so much time and effort on useless things such as art, music, poetry, and cracking jokes, he assumes that language was there all along. And it clearly wasn't. Yet, Miller finds it impossible to think of any other reason brain size would have increased. He fails to see any direct benefit to carrying around such an expensive organ. But surely it is caring for the helpless babies in our midst, whose care required our ancestors to build strong social relationships, that drove the need for language. It's the caring mind rather than the mating mind.

BACK TO MY IMAGINARY TRAIN JOURNEY

Obsessed as we are with our own evolutionary history, we continue to search for ancestors. Sometimes an unknown ancestor is found hiding within DNA. The genome of Denisovans contains 0.5 to 8 percent unknown DNA—not known from any other species of *Homo*. Where did it come from? Perhaps from a species that lived somewhere in modern-day Oceania and split from other species of *Homo* between 1.1 and 4 million years ago. That mystery relative could explain why Aboriginal Australians show genetic sequences in their genome not found in any other human population. The ancestors of indigenous Australians interbred with Denisovans and so could also have picked up some of the mystery DNA. There are probably more mundane explanations for the bits of unidentifiable DNA, but I bet that most of us much prefer the idea of a mystery species.

How many species could have been on my imaginary train? We'll never know, but more than the ones I started my train journey with at the beginning of this chapter. As recently as 2019, another new species was discovered on the island of Luzon, in the Philippines. It was given the name *Homo luzonensis*. Its remains were different enough from those of *H. floresiensis* to justify its own species name. Then there was a skull. That skull allegedly belonged to *Homo longi*, or Dragon Man. Dragon Man's skull was found in 1933 by a Chinese construction worker, who hid it in a well to keep it out of the hands of the Japanese occupiers. It wasn't described until 2021, after the construction worker, on his deathbed, told his family about the skull. Dragon Man had a big brain—as big a brain as our species. Its teeth were bigger than those of Neanderthals. Neither completely *Homo sapiens*, nor *Homo neanderthalensis*, nor any of the other known species of *Homo*, the authors made Dragon Man its own species. Dragon Man lived an estimated 150 thousand years ago. That makes it another potential co-commuter on my imaginary train ride, together with *Homo luzonensis*.

"Is 'Dragon Man' A Missing Link in Human Evolution?" one headline asked. That question reveals a particular way of thinking about evolution, a way that was common in the time of Marie Eugene Dubois, but should have been placed in the rarity cabinet of evolutionary thinking a long time ago. Evolution is not a hierarchical process. Yes, we still use the phrase "the tree of life," but the tree is a rather bushy one. Some of its branches are intertwined, where two species interbred, while other branches are short and lead nowhere. We don't know how many branches the human tree of life has—how many different species separate our last common ancestor and us. We probably never will know. I do know that one species did not simply become another species.

I once heard Richard Dawkins use a great phrase: "I didn't

wake up one morning being middle aged," he said. "Instead, I gradually changed from being young to middle aged." So it is with species. Our human tree contains perhaps hundreds of species, all connected with rather loose boundaries between them. A biological "link" may connect the many different species, but there is never just one species that is the missing link.

I think it sad that my train trip mentioned at the beginning of this chapter was only imaginary. How interesting would it have been for myself and my fellow *Homo sapiens* co-commuters to be surrounded by other species of human? But knowing my own species, we probably would not have treated those fellow travelers that well. Lacking our own ability for language, they would have seemed more akin to the other great apes than to us. And look how we are pushing them toward extinction. Language may have made us unique, but it did not necessarily make us kind. It did allow us one thing: to become master of our own destiny, to design our evolutionary future.

eight
BRAVE NEW WORLD

The 2016 science fiction film *Arrival*, adapted from *Story of Your Life* by Ted Chiang, features an alien language that allows those who learn it to (spoiler alert) see the future. It's a safe bet that precognition via linguistic innovation isn't the next step in human evolution. But what is? And what should it be?

In Aldous Huxley's 1932 dystopian science fiction novel *Brave New World*, The World State has created the happiest place possible by removing many human characteristics. No more annoying emotions. End to the individual. Instead, to paraphrase The World State's vision of "free love," "everyone belongs to everyone else." Children no longer have parents and are raised in artificial wombs in a Hatchery. There, the developing embryos are treated with precisely calibrated chemicals and hormones fitting them for their future roles in society. The Alphas get the best nutrition and grow up to be beautiful, ready to be allocated important roles. The Epsilons, the underdog working class, are given squirts of poison that stunts their development, rendering them able to perform the most menial and boring tasks without complaint.

The Hatchery is a paragon of efficiency. Embryos are divided, creating many clones. When fresh gametes are required, some of the Alphas are summoned to provide them. At other times, the Alpha females are required to take contraceptives, as unencumbered sex is strongly encouraged among Alphas as part of the happiness-making scheme. No sex for the other classes, though. They have work to do.

Lower castes are paid for their work in "soma," a hallucinogenic drug that keeps them calm. Alphas, too, regularly take soma when their frivolous lives become a little too meaningless. When the World State receives visitors from a "savage reservation," a place where normal people still live, both worlds are shocked. The "Utopians" are shocked by the uncouth habits of the savages. Imagine! The savages still practice marriage and monogamy, give birth to babies, and look after them. On top of all that, they are physically appalling. They grow old and look it, too. The savages, in turn, fail to see the appeal of a life without purpose. To them, science has done humanity an enormous disservice.

Aldous Huxley, the grandson of Thomas Huxley, Darwin's biggest defender, was not only a novelist but also a philosopher. In *Brave New World*, he attacked the commonly held Utopian belief of the time that progress in science and technology would provide the solution to all societal problems. Naïve perhaps, but that hasn't stopped us from trying.

Take Kazuo Ishiguro's 2021 novel *Klara and the Sun*. While Ishiguro's book is also a work of science fiction, it is more realistic than Huxley's. Klara is a humanoid, an Artificial Friend to lonely human children. Klara's charge, Josie, is sickly, but we are never directly told what has caused her sickness, nor the reasons behind the death of her older sister. Josie's friend, Rick, is an outcaste, maligned by other children and their parents. Halfway through the novel, it becomes clear that Josie, and her sister before her, have been genetically "enriched" to boost their academic performance. Rick's parents refused to have their son modified—a rare exception in a world in which almost all children are "improved" to give them a "fair" chance in life. Unfortunately, as it turns out, gene therapy—genetic enrichment—is not without risk, rendering Josie infirm and killing her sister.

How close are we to effective gene therapy in the real world? Traditional gene therapy uses viral vectors to transfer functional copies of genes into cultured cells that have been cloned from a human with a severe genetic disorder. By swapping the bad gene copies for good gene copies, the cells now start to produce the protein the patient needs to live a healthy life. The cells are then transferred back into the patient, where they chug away, making the life-saving protein in the patient. It's a great idea, especially in theory, but it doesn't always go well. In 1999, a teen died during a gene therapy trial that hoped to cure ornithine transcarbamylase deficiency, a rare metabolic disorder that results in the buildup of lethal levels of ammonia in the blood. Because of the use of a viral vector, a common side effect of gene therapy is a mild immune response against the virus. Sadly, in this instance, the teen had a very severe reaction. His body reacted so violently against the alien viral vector that he died. Ironically, the study was a safety study. The teen was a sufferer from the metabolic disorder, staying alive thanks to an inordinate number of pills per day. He had volunteered to participate in the study to see if the technique was safe before it would be developed further. Well, sadly, the answer was clear. It wasn't safe.

Understandably, the outcome of the study was a major setback for gene therapy, not to mention a devastating loss for the teen's family. But there were also other concerns. A major ethical concern of gene therapy is its potential to modify the germline of humans—the cells that produce eggs and sperm. While many people wouldn't mind fixing a defective gene in a fertilized egg, the technology becomes much more of a problem when the purpose of the "therapy" is to modify a human embryo that does not carry any heritable diseases. Should we start modifying embryos so that the resulting baby will grow to be more intelligent, as in *Klara and the Sun*? The ethical issues are so profound that almost all countries have banned making genetic modifications to the germline. If you genetically change a fertilized egg so that change will be inherited by its descendants, and you then implant that egg into a mother, you are breaking the law. That's a good law, I say, but it wasn't needed until recently because traditional gene therapy techniques were hit-and-miss (mostly miss) and could handle only one gene at a time.

Scientific progress has a way of getting ahead of the laws, and what was once impossible is now a reality. CRISPR/Cas9 biotechnology allows precisely targeted modifications to genes. Just cut out the bad bits and add a functional gene copy. It's a wonder cure that could be used to modify any fertilized egg. *Could* is now *has*. In 2022, Chinese biophysicist He Jiankui finished serving a three-year prison sentence for illegal medical practices. What was his crime? Altering genes in human embryos that were then brought to term after He forged medical documents and misled other doctors to do so. The goal was to make the babies—known by pseudonyms Lulu and Nana—less susceptible to HIV, a virus their father had unfortunately contracted. He Jiankui had used CRISPR/Cas9 to disable a gene that facilitates HIV infection. He Jiankui fell afoul of Chinese law because instead of "just" curing the babies after they were born (not that they needed curing, as they weren't sick), He Jiankui had changed the babies' germline, meaning the genetic change could potentially circulate in the human gene pool forever.

While we are probably not yet ready to irreversibly change the trajectory of our species' evolution by deliberately and irrevocably introducing "favorable" mutations using CRISPR/Cas9, we've taken baby steps in that direction. We can choose the sex of our babies by selective abortions or by only implanting IVF embryos of the "preferred" sex. Suffering from a heritable debilitating mitochondrial disease? Not a problem; we swap mom's faulty mitochondria for healthy donor mitochondria to make "three-parent babies." Ethicists, scientists, and medical practitioners are thinking hard about how decisions made by parents might ramify into future generations. In some countries, the individual decisions of parents have perturbed the natural numerical balance of boy babies and girl babies. Any evolutionary biologist, sociologist, criminologist, demographer, or economist worthy of their qualifications would have predicted things won't work out well if such a practice became widespread in a society. A highly skewed sex ratio, which always seems to be weighted toward males, will inevitably lead to some demographic and societal problems now playing out. Any society with a surplus of young men who can't find partners of their preferred sex is asking for trouble. But we have a long history of trying to manipulate our evolution. In the nineteenth century, it was scientists, not anxious parents, who led the charge.

In 1869, British polymath and the half-cousin of Charles Darwin, Francis Galton, wrote his book *Hereditary Genius*. Galton marshaled impressive data to argue that characteristics like intelligence, feeblemindedness, criminality, alcoholism, pauperism, schizophrenia, and manic depression were all highly heritable—and some of these traits assuredly are. Galton wanted to improve humanity by encouraging marriages between smart, sane, and pretty men and women. His movement—the eugenics movement—took its name from the Greek prefix *eu*, meaning "good," together with *genesis*, for the stock of raw materials in our collective gene pool. Darwin was not impressed by the eugenics movement. Not by its data, and not by the dubious morality of its social engineering, which in practice amounted to a thinly veiled effort to purge society of the "lower" classes. Wherever humanity suffered by nurture—or lack thereof—eugenics sought to rectify perceived deficits of nature.

Given the atrocities committed in the name of Galton's movement (see Carl Zimmer's *She Has Her Mother's Laugh* for a historical perspective), it is understandable that the idea of genetic differences underpinning traits like intelligence should elicit visceral, aversive responses to this day. Consider the complicated case of Edward O. Wilson, the late American biologist, naturalist, and popular author. In *Sociobiology: The New Synthesis*, Wilson explained the social structures of ant colonies—his favorite study organisms—and of other animal behavior using evolutionary theory and genetics. He might have evaded the ire of his colleagues and the broader public had he not included a final chapter on one especially salient social species: *Homo sapiens*. Days after Wilson's death in late December 2021, some blunt obituaries accused him of racism and sexism for his views on the genetic differences among us.

At the same time, people seem to have an enormous appetite for any information in their genetic blueprints that might shed light on observable characteristics and differences. Since the completion of the Human Genome Project, which kicked off the genomic revolution that gave us so much insight into our own history, we've been scouring the ledger for genes that code for X, Y, or Z—learning along the way that an awful lot gets lost in translation between genotype and phenotype. And as we've seen, CRISPR/Cas9 has opened the doors for manipulation of our source code that would have made a eugenicist blush.

To illustrate how far some people were willing to go in their quest to "improve" our species, I introduce the Soviet zoology professor Illia Ivanov. In the mid-1920s, Ivanov submitted a proposal to the Soviet government to inseminate humans with sperm from other great apes, and to inseminate chimpanzees with human sperm. At the time, it was all the rage among agricultural geneticists and breeders to make crosses between closely related species, with the idea of introducing "good" genes into one's species of interest. It seems likely that Ivanov followed the fashion of the time. The Bolsheviks were impressed and happy to fund the project. Off Ivanov went to French Guinea to catch some of the apes he needed. There, he inseminated female chimpanzees with human sperm, apparently obtained from his son. The reciprocal cross turned out to be a little more difficult. While Ivanov did not see any problems with inseminating women with sperm from one of the other great apes without their consent, the French governor, whose support Ivanov was reliant on, did not think this a good idea. Complaining to the Kremlin did not help, either. And so Ivanov had to find volunteers. Which, amazingly, he did, although the experiment was never carried out because Ivanov was arrested and sent into exile in Kazakhstan for reasons that had nothing to do with his "science." No further mention was made of the fate of the inseminated female chimpanzees, but it is safe to assume that no pregnancy resulted from this appalling "experiment," not least because humans and chimpanzees do not even have the same number of chromosomes.

How did we become so obsessed with our species' destiny?

Language. Without language, there can be no science, and without science, there is no technology. Science and technology allow us to fast-forward our own evolutionary trajectory. But at what price?

At the end of Stanley Kubrick's epic 1968 film *2001: A Space Odyssey*, the single survivor of the spacecraft *Discovery One* ponders the eclipse of human intelligence. HAL, the heuristically programmed algorithmic 9000 computer designed to maintain the ship's functions and to communicate with the crew, has gone rogue, killing all humans on board bar one. Once the crew realized HAL's evil intent, it was too late to disconnect the computer before it went on its killing spree. What the crew failed to realize was that HAL had taught itself to lip-read so it could thwart every attempt of the crew to switch it off. In one of the film's final scenes, the last remaining human turns off HAL by unscrewing his memory cards one by one, while the machine begs him to change his mind. With each unscrewed card, HAL loses some of its functions. The last turn of the last screw renders the computer impotent.

Kubrick was way ahead of his time. Where will we find the screws to undo the power of self-replicating AI? Perhaps the cure will be found by scientists working at the University of Cambridge's Centre for the Study of Existential Risk. The Centre is inspirationally situated adjacent to the Mathematical Bridge. This bridge, built in 1749, is a tribute to the brilliance of mid-eighteenth-century engineering. Some claim that Sir Isaac Newton himself designed the bridge to illustrate the explanatory power of physics. If so, he could have potentially illustrated a lot more about (meta)physics, as he had died twenty-two years before the bridge was constructed. Anyway, the Centre's raison d'être is to study possible extinction-level threats posed by present and future technology. Guess what is one of the Centre's main research priorities? Existential risks from artificial intelligence.

Language got us into this quagmire; it is also the only tool that can get us out of it.

THE MORAL APE

My eldest daughter works at a center for the criminally insane in Utrecht, the Netherlands. All inmates—referred to as patients—have committed crimes for which they went to prison. After their prison time, they move to the clinic; they are "placed under a hospital order." The UK equivalent term for this is "to remain at His Majesty's pleasure," but whatever the term, the patients remain in the clinic until the day they are deemed safe to return to society. Many will indeed make it back into society. Others do not, unable to overcome whatever urges made them commit their crime.

Despite these patients' criminal histories, staff at the clinic are not armed, nor do they carry cell phones while on duty. Evening meals are prepared by teams of patients and shared with their supervisors, who often are females in their twenties or early thirties. Picture the situation. Groups of about seven or eight former mostly, but not exclusively, male prisoners, all with mental health issues, many convicted of crimes of a sexual nature, in the same room as an unarmed young female. Like any ordinary family, they enjoy their meal, each other's company, and chat about their day.

Why doesn't the situation escalate? After all, these are not your ordinary dinner companions. "Simple," my daughter explained when I asked. The patients have been given the responsibility to keep everyone safe. They make sure edgy situations do not escalate. Attuned to subtle changes in mood or behavior, the patients are adept at putting the genie back in the bottle. Most of the time it works.

My other daughter works for a major bank, also in Utrecht, in the Netherlands, in the fraud department. That name really should be the "anti-fraud department" because her role is to recognize fraudulent transactions, to investigate, and if necessary to intervene. In quite a few instances the people she investigates are her own colleagues. Clearly, for some people the temptation is too strong. Not satisfied with a well-paid job, they find the easy access to others' financial details hard to resist. During the interviews that are part of an investigation, my daughter often hears the most fantastical explanations for the alleged fraud. They did it to help someone. It's self-deception in action.

Why can former criminals become keepers of the peace while otherwise law-abiding citizens can't resist squirreling away money from customers who have placed their trust in them?

Morality has something to with it, including the ability to distinguish between what is right and what is wrong. The former criminals know that they, too, will get into trouble if one of them goes into a rage. The fraudulent bank employees live in a fantasy world, one in which they justify their actions by believing they are doing it for others. Their sense of morality is warped.

Morality. Have we finally found something uniquely human? Before you answer that question, let me tell you about an experiment on capuchin monkeys, performed by Frans de Waal in collaboration with Sarah Brosnan. Over his long career working on apes and monkeys, de Waal has shown us how our emotions are similar to those of our nonhuman relatives. Refusing to see humans as something special—a species somehow outside of nature—de Waal's work points to the origin of tendencies we too often claim to be uniquely human. Our emotions did not come out of the blue, though; our relatives have them, too. Take fairness. Fairness is an essential component of morality, and so de Waal and Brosnan looked at capuchin monkeys to see if they have a sense of fairness.

Two monkeys—members of the same group—were placed in adjacent test chambers separated by a mesh partition and were trained to return a token for a reward. One at a time, they were given a token; the experimenter would then hold out their hand and if the monkey returned the token, it would get a piece of cucumber. While it would happily munch on the cucumber, the other monkey was performing the same action. All went well until one of the monkeys was given a grape as reward instead of a piece of cucumber. Although cucumber is perfectly acceptable, monkeys regard grapes as much yummier than cucumbers.

The first time one of the monkeys saw its partner getting

a better reward for the same task, it dealt with it, grudgingly. (I highly recommend watching de Waal's TED talk which has a video of the experiment. The monkey's reaction is truly hilarious.) But the second time the grape injustice took place, it broke into a rage, chucked the cucumber at the experimenter, and rattled the cage in frustration. Clearly, life is not fair and this monkey knew it.

What exactly the furious monkey thought as it refused its piece of cucumber we can only guess. We can try and ask it, but the monkey can't answer us. So, another set of researchers came up with a clever experiment designed to determine the thought processes that underlaid the cucumber refusal. Their animal of choice was another species of monkey, the long-tailed macaque. In their experiment, the macaques were either tested alone or, as in the capuchin monkey experiment, with a partner. When a macaque was offered a food reward of lower quality by an actual human, both with and without another macaque present, it was much more likely to refuse the food offered than when the reward was delivered by a machine. Instead of feeling that they were being treated unfairly, it seems the macaques suffered from social disappointment. One suffers from "social disappointment" when one is not treated as one expects to be treated. Clearly, the machine was less to blame for the macaque's appalling treatment than an actual human was. Or, so went the thinking of the researchers.

We are quite like macaques, it seems. At least that is what the Ultimatum Game suggests, a beloved tool of sociologists. The idea is simple. Two players, one choice. The choice is how to split a certain amount of money, say $20. The first player decides how to split the money. Any offer is allowed, but the player cannot keep the whole amount. The $20 needs to be split, and a split of $1 to $19, say, is permissible. The second player then has a choice: accept the offer or reject it. Rejecting the offer means that neither player gets any money—and that never makes any sense economically. No matter how unfair, a rational being should accept any offer because the alternative is not getting anything at all.

Well, that is not what happens. Unfair offers, in which the first player allocates significantly more than half the amount to themselves, are very likely to be rejected. Such offers are refused even when the offer is made by a computer instead of a fellow human being, although the sense of unfairness is less in the case of a computer. It appears that humans, capuchin monkeys, long-tailed macaques, and undoubtedly many other species are prepared to forgo a small reward in order to communicate that they have suffered unfair treatment. That emotion is so strong it can't be suppressed completely, even when the injustice is inflicted by a machine.

Almost everything we regard as being "uniquely" human, such as our ability to distinguish right from wrong, has a long history in the simian lineage. Morality did not, as most humanists and utilitarian philosophers contend, arise from our uniquely human ability to use reason to figure out better ways of structuring our societies and personal relationships. Nor is morality a human invention explicitly devised to control our natural tendencies to be selfish and horrible—the view of seventeenth-century English philosopher Thomas Hobbes. *Morality* is a loaded term, but in essence it comes down to notions like fairness, abiding by social norms, and the ability to feel empathy.

All social animals need a set of rules to live by, otherwise their society fails—even simple critters like social insects. Many species have evolved a kind of police force—individuals that punish those that try to cheat. (Cheating in the social-insect world mostly relates to females other than the reigning queen trying to lay eggs; remember what happened to the subordinate meerkat females that tried to make their own babies.) Dinosaur ants, so called because they are rather large for an ant at over 1 inch long, forcefully pin down nest mates trying to lay eggs. Honeybee workers also punish individuals trying to lay eggs. They check every egg laid to see if they carry the "royal seal," a chemical mark deposited on the egg by the queen that workers cannot fake. If the chemical mark is absent, the police workers eat the offending egg because no mark means that a worker is trying to cheat. Insect colonies, like meerkat societies, can function well only if reproduction is strictly regulated.

Of course, I am not arguing the insects have a sense of morality. What I am saying is that emotions underlying many aspects of morality have a natural basis and are not a human invention. When a parent is tuned into the needs of their newborn, we call it *empathy*, the ability to understand the needs and feel the emotions of the other. If a nonhuman animal shows the same attention to its newborn, we are likely to dismiss it as instinct. All our emotions have their antecedents in our evolutionary history. As the Australian moral philosopher Peter Singer argues, we can use rationality, reason, morality, and ethics to rise above our evolutionary destiny. If we so choose. Language gave us that ability. It is up to us what we do with it.

MACHINES LIKE US

This book could have been written by ChatGPT. Well, maybe not quite. I could certainly have used the chatbot to make my life easier. I typed in "Write something under the header 'Machines Like Us,'" and the chatbot gave me 424 words in a few seconds. Too many words to reproduce here, but the last paragraph is worth quoting in full:

> In conclusion, the emergence of machines like us marks an exciting era in human history. While we must address the ethical implications and challenges that arise, the potential benefits are immense. By embracing AI technology responsibly and fostering collaboration between humans and machines, we can unlock new possibilities and shape a future where AI enhances our lives in meaningful and profound ways.

Note the word *responsibly*. There is some irony in a machine warning us against the use of machines, isn't there?

According to some influential people—those affiliated with the Centre for the Study of Existential Risk for starters, including prominent scientists such as the late Professor Stephen Hawking—the main threat coming from generative AI, that jack-of-all-trades AI, is that it may become conscious and develop a morality differing from our own. That different morality may see the end of us if the conscious AI decides humans are a bit of a bother and the world would be a better place without us—not a totally unreasonable conjecture. I personally do not think the danger of AI lies in its potential to become conscious, for the simple reason I do not think anything not alive can become conscious (which is not the same as saying everything alive is conscious). To my mind the dangers are much more mundane.

Take my chatGPT text as an example. When I asked the chatbot my question, it trolled the internet to find related content and strung that content together, using the language rules the algorithm was trained on, to come up with its suggestions. But what it did not do, and in fact cannot do, is check the veracity of the information it found on the internet. If social media are anything to go by, false information can spread on the internet like wildfire. Then, there are issues with copyrights and plagiarism, badly trained algorithms, the inability to easily fact-check, and myriad other problems.

Despite these many concerns, the search for applications of AI continues unabated. Take the phenomenon known as the "hungry judge effect"—the notion that judges are likely to make different judgments before and after lunch. If a human judge's judgment is so dependent on their physiological state, wouldn't it be better to simply let an algorithm decide the defendant's fate? Such algorithms could, in theory, quickly and efficiently gather all similar cases, their outcomes, and based on precedent, decide the case before it. That's assuming, of course, we have figured out how to ensure the information so gathered is reliable. I can see that a judge can use that information to come to their own judgment, but I would hope we never have a situation in which human judges have been made redundant.

Perhaps the horse has already left the stable. To my dismay, I found a disturbing suggestion in the New South Wales, Australia, Handbook for Judicial Officers. Under the heading "Technology Supporting Judges," I found a mention of transhumanism, aimed at "improving the human condition through applied reason, especially by developing and making widely available technologies to eliminate aging and to greatly enhance human intellectual, physical, and psychological capacities." Someone is clearly thinking big, but what follows is scary. The handbook suggests that judges of the future should interphase their brains directly with computer memory and AI to "increase their intelligence and memory, increase their ability to manage and process information, and reduce the occurrence of fatigue." You wish. Actually, I think no one should wish anything like it.

Instead of making human-robot cyborgs, it is much more likely we design machines that look and act like us. Machines like Ava, from the 2014 film *Ex Machina*. Ava doesn't quite have the body of a human, but she has a human face and the general physical features of a gorgeous female. But above all, she speaks like a human. Like a female human, tricking the human male, Caleb, into believing she is conscious and she is in danger of being changed by her designer. Playing on Caleb's emotions, Ava promises she will run away with him if he helps her escape. Predictably, things do not end well for Caleb; he is murdered by another cyborg, one mistreated by its designer, and Ava enters the real world, indistinguishable from a human.

The film cleverly shows how important language is to us humans. Ava didn't have to have the exact body of a human (before she escapes, Ava dresses to disguise her robot body); she just had to be human enough. Caleb knew he was falling in love with a machine, but Ava's ability to read his mind, understand his feelings, and react to subtle nonverbal hints made him ignore her true nature. Instead, he empathizes with Ava as he would with a real woman.

I think we very likely will treat machines like us as we treat each other. If they speak like us, that is. When you think about it, that would be rather bizarre. Consider the way we tend to treat nonhuman animals as lesser beings. Or, people less than fluent in our own language. The knee-jerk tendency to think of them as being of lower intelligence is hard to avoid. Given the value we attach to language, my bet is that machines created in our *speaking* image will most likely be treated better that our nonspeaking relatives.

Enough gazing into the crystal ball. It's time to explore how exactly language allowed us to take control of our destiny—not by changing us directly à la Galton, but by changing our environment.

FROM CAVEMAN TO THE MOON IN NO TIME

My mother's grandfather—my great-grandfather—was born in 1885. As a child, I was amazed that someone that old could even be alive. Here was a man who, when young, knew only horse-drawn carriages, saw the first car arrive in the Netherlands (Karl Benz's invention of the Patent Motorwagen on May 19, 1896), remembered the invention of the diesel engine (1893) and wireless communication (1895), and the coming of radio broadcasts, and who was much more cognizant than I was when Neil Armstrong set the first foot on the moon on July 20, 1969. Not only had technology advanced at a rate never seen before, but medical innovations did too—to the point that the population of the Netherlands grew from 4.27 million at the time of my great-grandfather's birth to a whooping 14.22 million when he died. The country itself grew as well, thanks to the success by the Dutch in reclaiming land from the sea.

How did our species manage to morph from caveman to space explorer in a blink of an eye? The answer is, of course, language. Language changed us quicker than natural selection ever could have because it allowed humans to cooperate in a totally new way. Language allowed us to live in large groups, held together by stories, gossip, shared belief systems, art, music, and what else have you. Instead of "wise man," the meaning of *Homo sapiens*, we should rename ourselves *Homo largiloquus*, "talkative man." That's not because our species alone has a special gift for language, thanks to a language-mutation à la Chomsky, but because we have made the best of our linguistic ability to learn—and to learn fast. Initially, we learned to use language to better raise our children so we could produce more of them by becoming hyper-social. That hyper-social environment then allowed us to start molding our own future by making it even easier to learn, to make ever more sophisticated tools. Language sped up our evolution, at the exclusion of all other species of *Homo*.

MAN IS BUT A WORM

After Darwin published his evolutionary theory, many made fun of his thesis, particularly where it related to us. The cartoonist (Edward) Linley Sambourne drew a cartoon of a worm slowly turning into Darwin, captioning it "Man is but a worm," playing on the title of yet another book by Darwin, *The Formation of Vegetable Mould through the Action of Worms, with*

Observations of their Habits, published in 1881. I, too, learned something from worms. As so often happens, that occurred quite serendipitously, as I sat next to someone in a pub who studied worms.

My pubmate's worms, a species of roundworm or nematode, came in two forms: those with a large mouth and those with a small mouth. Not surprisingly, what mouth size they had affected what they could eat. Mouth size is not influenced by genes directly; there is no gene for small mouth or large mouth, but rather by epigenetic marks that affect gene expression. The environment, then, has an influence on the epigenetic marks, so that the worms, during their development, adapt to the size of prey they encounter most. Lots of large prey developed a big mouth; mostly small prey kept it small. It's a great example of *phenotypic plasticity*. Phenotypic plasticity is a well-known biological phenomenon; recall the plant examples mentioned in Chapter Two—my favorite, the master mimic vine that changes its leaves depending on where it grows—but what I found the most interesting is what happened next in their evolutionary process.

If the worms found themselves in a stable environment where prey size doesn't change over a couple of generations, the worms lose the ability to form the alternative mouth size. So, a stable small prey environment led to worms that no longer could develop a large mouth, and vice versa. The worms' phenotype became *canalized*. Where initially all worms could hedge their bets on what environment they would find themselves in, now only one developmental outcome was possible. No more phenotypic plasticity. The evolutionary biologist Mary Jane West-Eberhard called this phenomenon "plasticity first" evolution. Epigenetics allows phenotypic plasticity, which then becomes fixed if the environment no longer changes, so now the organism is well adapted to that particular environment. This rather difficult-to-grasp concept is nicely explained in Ben Oldroyd's book *Beyond DNA*. A behavioral geneticist, Oldroyd provides many examples of instances where an organism's phenotype first adapts to its prevailing environment before the trait becomes encoded into DNA.

What do these worms have to do with us? They illustrate how plasticity can allow a species to quickly adapt to its environment without having to wait for the right mutation or mutations to pop up. And we need to find some quick mechanisms to explain how our young species managed to become the dominant species in so little time. One key to our success was the ability to adapt to new environments, while our relatives stayed behind in the trees.

In 1896, the American psychologist James Baldwin hypothesized that learning is a means to speed up evolution. Instead of adapting to new environments through the accumulation of new mutations, which is a slow process, learning would allow a species to quickly settle into new environments. Just as the worms, through their plasticity, could adapt to different environments, with big and small prey, the ability to learn gives a species a lot of flexibility without such flexibility being coded into the DNA. (Of course, the ability to learn would have some genetic basis, but one does not need to infer some gene "for" learning.) Individuals that learn to adapt do better, and so the species, over time, becomes even better at learning. That increased ability to learn then becomes fixed in the population because the set of genes, or the particular ways in which genes are expressed, are linked to learning and become fixed owing to directional selection on the better learners.

The so-called Baldwin effect is definitively not generally accepted by evolutionary biologists. While it is easy to verbally argue that plasticity can speed up evolutionary change, and that such plasticity ultimately leads to genetic changes, actual examples of the effect are sparse. The worms of the colleague I caught up with in the pub could well be one of the first clear examples of the role of plasticity on genetic change, provided my colleague's team can pinpoint the associated genetic changes—something they are working on. Whatever label one prefers to attach, learning definitely allows for a level of flexibility that was new in the history of life on earth.

Perhaps one reason why many evolutionary biologists remain unconvinced about the Baldwin effect is that it has been evoked to explain the origin of language and cognition without really giving a satisfying mechanistic explanation. We now know there is no such thing as *the* language gene, making it difficult to see how our initially primitive ability to use language could have led to specific genetic changes as per the Baldwin effect. I won't attempt to give a satisfying explanation because I prefer to turn matters around. In my mind, once we acquired language, our learning abilities accelerated because we could more effectively teach others and make use of the knowledge stored in many minds. We created the perfect environment for ourselves, and so our species blossomed. But as always, we merely expanded an existing tendency, building on the foundations that had been established before our ancestors split from the ancestors of the other great apes.

CURIOSITY KILLED THE CAT. OR DID IT?

Without at least some level of curiosity, one would never discover a new food source. And without curiosity, and bravery, one would never learn if the new food source is palatable or toxic. But as Eve found out after she gave in to the temptation to eat the forbidden fruit, curiosity can have negative consequences. So, how does one balance the two, exploring new territory with a potential for reward and risking getting hurt or even dying in the process? By taking cues from others.

Birds do it all the time. Seeing another bird feeding is a good indication there is something to be found; better check it out before heading off on your own. When a bird returns with food, it seems a sensible strategy to follow it on its next venture out. To take cues from others, one doesn't need to live in some sort of stable social group. One just needs to be around others sufficiently to be able to observe what they are doing. In a way, the ignorant individuals can parasitize the knowledge of others. In fact, lots of species use the eyes and ears of other species. Take hyenas; they observe circling vultures to identify kill sites (as do safari guides). You might say they cheat. Does that mean when you do live in a social group, you are even more likely to cheat? Or, to put it into kinder terms, are social species more or less likely to be curious, finding things out by themselves using trial and error? That is just what a recent study wanted to find out, using chimpanzees, bonobos, Sumatran orangutans, and Bornean orangutans as test species.

Individuals from each species were given the opportunity to discover a food item they had never seen before. That they had never seen it before was not surprising, given the food was a potato mash formed into a ball colored blue with black olives. As a measure of curiosity, the researchers used the length of time it took each individual to first touch the weird item and then, if they were brave enough to try it, to first have a taste. The quicker they touched and tasted the ball, the more curious they were deemed to be.

The apes were also tested on their reaction to a never-before-seen toy, made of different colored tennis balls that were loosely bolted onto a wooden plate, so the balls could be moved sideways and rotated. Would they or wouldn't they start to try to figure out what the thing was for? Again, the quicker they approached the weird construction to figure out how it worked, the higher their level of curiosity was deemed to be. In another challenge, an unknown person dressed in black would appear in front of the apes' enclosure. The apes could either show signs of curiosity, moving closer to the stranger and trying to engage or be totally uninterested (all apes were habituated to humans, so they were never scared even of unknown persons).

As mentioned, when living in a stable social group, a species could reduce its chances of getting into trouble simply by observing what others do. If they eat something you have never seen before, it's a good bet the food is good to eat. If a weird object is being played with, then why wouldn't you do that too? In other words, you need to be less curious when always surrounded by others whose behavior you can observe. That idea—that the more social species are *less* curious than the more solitary species—is known as the *social curiosity hypothesis*. In this experiment, the participating chimpanzees, bonobos, Sumatran orangutans, and Bornean orangutans all played their game well. The species that live in the largest, stable groups with the highest tolerance of strangers—the bonobos—were the least curious. The most solitary species with low levels of tolerance of others—the Bornean orangutan—turned out to be the most curious. The other two species fell in between, with chimpanzees being less curious than the Sumatran orangutan. All the species scored equally well on cognitive tests, so one cannot explain the lower tendency for curiosity as a lack of understanding.

I like the social curiosity hypothesis. Given that finding things out always has some level of risk, any means by which you can reduce that risk is a bonus. It then follows that when living in a group of trustworthy individuals, one doesn't need to figure things out for oneself. Just do as others do and you'll be fine. Now, imagine the additional power that the ability to communicate would give a species. Instead of having to keep a keen eye on what others are doing and learning from them by observation, with language one can simply tell others what food is good to eat and how to prepare it, what scary-looking animal one needs to be careful of and which one is innocent even though it doesn't look it. The downside is having to tolerate the insatiable barrage of questions that human infants tend to throw at adults. I'd say that is a small price to pay.

ANCIENT INFORMATION

In 1900, while sheltering from a storm near the tiny island of Antikythera, between Crete and mainland Greece, sponge divers found more than they bargained for: a shipwreck containing a wealth of Greek artifacts, including many marble statues. In the midst of all the excitement, one retrieved object was initially ignored. It was a lump of who-knows-what the size of a large dictionary. While in storage at the National Archeological Museum in Athens, the lump broke apart, revealing objects seemingly impossible for the time: precision gearwheels the size of coins, cast from bronze. Impossible perhaps, but here they were.

The gearwheels were part of what is now known as the Antikythera mechanism. It took scientists more than 120 years to understand the machine's function. Referred to as the world's first mechanical computer, the Antikythera mechanism was an intricate device that tracked the position of the planets as viewed from Greece. With the Antikythera mechanism, the ancient Greeks could predict the position of the sun, moon, and planets on any specific day in the past or future. By this means, the constructors of the machine could have predicted major celestial events such as lunar and solar eclipses. The Antikythera mechanism was constructed between 200 and 60 BCE, at a time during which the Greeks still thought the earth was the center of the universe.

In constructing the Antikythera mechanism, the ancient Greeks built on knowledge from Babylonian astronomy, mathematics from Plato's Academy, and ancient Greek astronomical theories, based on their nightly observations of celestial bodies using their naked eyes. They succeeded because they could build on accumulated knowledge. This is the power of language: the unrivaled ability to effectively transmit information.

Our ancestors used language to reduce their dependence on trial and error in exploring and exploiting new food sources, thereby allowing them to expand the kind of environments they could thrive in. We, through effective communication, became specialists in overcoming the natural defenses of plants and animals that evolved to reduce their risk of being consumed. Now, only one or a few individuals had to figure out how to detoxify plants or prevent being killed by an animal. A successful hunter, one who learned the movements of their favorite prey species, could, through language, teach youngsters which tracks to follow and what the animal's most likely movement patterns would be, given, say, the time of day or season. Language gave us the ability to reason, to come up with intuitive theories about our surroundings. Reason allowed our species to learn to exploit other animals and plants before they could evolve counter-mechanisms.

If language has led to so much that made us uniquely human, how can I be so sure that language evolved first and foremost to assist in the rearing of children? To answer that question, it is important to make a distinction between what language is used for and what it was adaptive for. To understand *why* language evolved, we need to understand how language gave our young species an advantage in the evolutionary arms race.

Darwin's theory of evolution through natural selection boils down to a simple idea: In each generation, more offspring are produced than can survive and reproduce. If a little bit of language helped one family do better than another in keeping their children alive and letting them prosper, then not only would language be favored by natural selection but it would also be under selective pressure to become more sophisticated—and an even better help for raising a family. To appropriate a metaphor Peter Singer used, getting from reason to ethics (beginning to use language because of our childcare needs) is like stepping onto an escalator that leads upward and out of sight. Once our species placed a first foot on that language escalator, we could only go up, broadening the use of language far beyond childcare. This is so much so that language allowed us not just to understand the universe we live in but also to exert control over our evolutionary trajectory.

And this brings me to an often-posed question: Are we still evolving? Have we taken so much control over our destiny that biology has become irrelevant? The Flynn effect certainly seems to point that way. Named after the American-born New Zealand researcher James Flynn, this effect describes the steady increase in IQ scores since the early 1930s, not just in the United States or New Zealand but globally. If my great-grandfather would still be alive today, I would be much smarter than he was. Or, at least I would do better at the tests used to measure IQ.

Such a rapid increase in IQ over just a few generations cannot be explained by natural selection; it has simply been too quick. The rise in IQ is probably due to a range of factors: familiarity with the kinds of questions asked on IQ tests, better nutrition and physical health, and a reduction in commercial products that contain lead (known to induce myriad neurological disorders). Also, the world has become a more complex place, one in which technology plays an ever-increasing role. I therefore have to navigate a significantly different world than my great-grandfather had to. But what is weird is that the Flynn effect has reversed since its peak in the 1990s in countries like Norway, Denmark, Australia, Britain, the Netherlands, Sweden, Finland, and the United States. Perhaps it is the reliance on certain kinds of technology that is now becoming a problem. Whatever the reason for the reversed effect, it seems a safe bet that my grandchildren won't be smarter than I am. I am not quite sure if that is a relief or a disappointment.

As we have modified our environment, we have modified ourselves. With the advent of agriculture, we started to live in much larger groups, ultimately building cities, creating the need to become even more cooperative, even with total strangers. Such artificial selection is nothing special; humans have been artificially selecting plants and animals for thousands of years. What is special is that we are the ones causing our own selection. Modern technologies, such as reproductive advances, certainly take some of the power out of the hands of natural selection. IVF and other assisted-reproductive technologies allow couples to reproduce who would not otherwise be able to conceive naturally. With baby formula available, every baby can thrive irrespective of the mother's milk production. And thanks to superb healthcare, corrective lenses, false teeth, and hearing aids, my great-grandfather lived in reasonable health to the ripe old age of ninety-six. Just think about the additional children he could have sired during his technology-supported long life if he had so wished. (My great-grandfather still had a respectable four children.)

While technology may affect the direction of our evolutionary trajectory, natural selection remains a powerful force. We can see the effect of evolution through natural selection in changes in our genetic makeup. Despite the extremely high genetic similarity of all humans alive today, we can still find differences that, although small, have significant effects. We can trace the fate of the lactase gene, a gene that helps us digest the lactose sugars in milk. When some populations in Africa and the Middle East began to raise cattle and drink milk as adults to supplement a poor diet, the lactase gene mutated so it no longer switched off after weaning, the normal course of events in mammals. Why continue to make an expensive protein when it is no longer needed? A functioning lactase gene persists in adults descended from these populations and allows them to digest milk. Lactose intolerance is therefore not an aberration—the ability to digest milk as an adult is. To offer another dietary example, many Greenlandic Inuits have inherited a particular gene variant that prevents them from developing the negative effects of a high-fat diet.

Our past dalliances with our close relatives, the Neanderthals and Denisovans, have also left their calling cards in our DNA. By the time our species made it to Europe and Asia, Neanderthals and Denisovans had already adapted to the local climatic conditions. By mating with their relatives, our ancestors could quickly pick up genes that helped them adapt to local climatic conditions, without having to wait millennia for natural selection to favor their own versions of genes to survive these new environments. People native to the high-altitude Tibetan Plateau carry in their genome unique genetic variants that allow them to live in a low-oxygen environment. The gene variants are leftovers from past interbreeding events with Denisovans.

As always, though, nothing comes for free. There are also downsides to our species' interspecific sexual adventures. When SARS-CoV-2 hit our species, not everyone was equally likely to develop severe COVID-19 symptoms. People who inherited a stretch of Neanderthal DNA that resides on chromosome 3 were particularly at risk. This stretch of DNA increased in frequency since the Last Glacial Maximum—between 31,000 and 16,000 years ago—probably because it protected against prevailing diseases at the time. Now it is found in 16 percent and 50 percent of people in Europe and Asia, respectively, suggesting a strong benefit to carrying that bit of DNA. One benefit, as it turns out, is a 27 percent reduction in the risk of becoming infected with HIV. The trouble is that the same mechanism that prevents an HIV infection makes it easier for SARS-CoV-2 to infect cells.

In the end, despite all our achievements, including vaccines, even we cannot rise above our biology.

A UNIQUE RESPONSIBILITY

Yes, we are a unique species, but so are all other species. We may have language, but we can't fly unassisted, can see only a small range of colors, our sense of smell is the worst of all mammals, and let's not mention our inability to detect the earth's magnetic field or to use echolocation. Despite these deficiencies, our linguistic abilities mean our species inhabits a unique space in the natural world. We are the sole survivor of a lineage that spanned millions of years. Our species exerts unprecedented control over all other living creatures. Language gave us the power to dream, to think, to plan, to do. And to argue that we are something different from all other life forms on earth, proudly so. In some sense we are different, and maybe to be proud isn't so bad. But let us remember that with our unique place in nature comes a unique responsibility: to care for all other unique species with whom we still share this planet.

After all, that is where it all started: with the language of care.

epilogue
MODERN FAMILY

To breed or not to breed? Sometimes this can seem an agonizing choice between life's realities and life's expectations. People who don't want to be parents worry about bucking their biological imperative and being seen as selfish. Others, who want to be parents someday, are already overextended, concerned about their ability to make time for their future children in the way our society makes us feel we should.

These dilemmas are all modern constructions. None of us were "meant" to go it alone. Nature didn't design us to care for self and child without a wealth of human resources. If it feels like the parenting job is overwhelming—even if you have support and only solo-parent from time to time—I hope you'll remember not to blame yourself. Especially if you're a *mother*.

Western societies have internalized the narrative that mothers have some special instinct for child nurture that comes online in motherhood. That we are natural-born caregivers, capable of single-handedly giving our children all the love and support they need to thrive. There is even a name for it: mother instinct.

In her book *Mother Brain*, science journalist Chelsea Conaboy dispels the myth of the mother instinct, the notion that motherhood is hardwired in women, leading to instant ecstasy as soon as the baby is born. How did we get ourselves into this pickle?

Maybe we should blame the industrial revolution. Before we became slaves to machines, most people lived on farms where family members, old and young, great and small, contributed to the welfare of the extended family. Farm work is tough, but farm life itself had many benefits, especially for mothers and babies. No matter how young, babies could come along to the field or shed, strapped to anyone big enough to carry them for a while. Mom would be there to feed the babe when needed, but she didn't have to bear sole responsibility for looking after the baby. Siblings, father, and grandparents were typically on hand. Childcare was a collective enterprise, as it had been for millennia. Even with the advent of cottage industries, like weaving, leather manufacturing, and other handicrafts, it was all centered on the family home; the family worked as one, sharing tasks. It was family as a collective enterprise. But that all changed when industrialized factories made their appearance, and work and family became separated.

Designed to house machines too large and expensive for any one family to own at home, factories became paragons of efficiency, subjugating the individual worker to the service of the factory owner. (Charlie Chaplin's classic 1936 film *Modern Times* comes to mind.) Industrialization would eventually create the conditions for the birth of the nuclear family, with mom's main job being to stay home with the kids.

If you read this far, you will have come to realize that the nuclear family—dad who worked outside the home to provide for the family, mom whose responsibilities were confined to the home, is not the kind of family structure that enabled our species to prosper in the last few hundred thousand years. The "traditional" family favored by many conservatives today is far from traditional. Rather, it is a relatively recent social construct, mainly in Western societies, that has been given tendentious scientific support via a peculiar interpretation of our evolutionary history by some evolutionary psychologists and sociologists. The nuclear family goes *against* our nature. We are the most social species of all mammals. Our need for social contact is the reason solitary confinement is often regarded as one of the worst forms of torture for us; we can't stand being alone. Who could have possibly thought that raising kids in isolation is good for them? Or for the parent?

Obviously, I would be the last to argue that biology has nothing to say about which parent is most likely to feel most responsible for the baby. It is the female of the species who is pregnant, gives birth, and breast-feeds. These are all important facts that affect the psychological makeup of females. As a mother myself, I can attest to the hormonal changes that surge through the body during pregnancy and childbirth, and the effects that a baby has on female emotions. But female biology does not mean that human mothers are supposed to be the sole caregivers to their offspring.

Cultural evolution brought us the nuclear family. Most of the time, we see cultural evolution as a positive force, trumping biological evolution in the speed with which it can effect positive change. But the emergence of the nuclear family was far from an unalloyed good, and it has pushed us too far from our biological roots. Thanks to language, our species could raise more children by recruiting helpers. As modern parents become more isolated from their extended family, raising the next generation becomes an even bigger challenge. Many children in the Western world grow up without grandparents, either because they no longer live nearby or they died owing to the later age at which many women now become mothers. I had a fortunate childhood. Not only did I have my grandparents, but I also had my great-grandparents until I was well into my teens. They were of great help to my parents, emotionally and, when the need arose, financially. For me, they were repositories of knowledge, my emotional soundboards. They spoiled me, too, knowing well that my parents would have to deal with the consequences. Grandparents and other elders have played such roles throughout most of human history and prehistory.

There is a danger we are losing something precious. Something uniquely human. The perfect conditions to learn language.

◆

Meet Sam, a baby boy who lives near Adelaide, the capital of South Australia. From the age of six months until he was about two years old, Sam wore a helmet with a mounted camera. Not all the time, of course, but for about an hour twice a week. Sam took part in a research project that aimed at understanding how children learn language so that apps like ChatGPT can become better at fooling us. Or helping us, depending on how it is used.

The visual and auditory information gathered from Sam's camera was used to build a Child's View for Contrastive Learning (CVCL) model, a model that mimicked how Sam seemed to associate objects with sounds, to construct words. Just like Sam, the learning model updated its information with each experience, with no other input than the information that came from Sam's camera. (In computer science jargon, the model trained in a self-supervised manner.) Every time a particular object—say, a ball—was seen in combination with the same utterance, "ball," the training model's association between the visual and auditory information was strengthened. The word *ball* became associated with the object ball.

Having so trained the CVCL model, based on Sam's experiences, the learning model was then put to the test. Could it correctly associate pictures of objects with words? It could. It could even generalize beyond the exact words that it learned via Sam. A key to its success, and that of Sam, was that the same or similar object—a ball—was seen in different contexts. Sometimes a ball was found at the bottom of a staircase, other times in the sand pit. A ball could be a basketball or a yellow squeezy ball. Balls could be red or blue, or whatever other color, but the word would always refer to a round object that could be picked up and played with. And so both the algorithm and Sam learned the generalized concept of what a "ball" is, instead of one particular manifestation. (Each time you are asked by some website to prove you are human, not a robot, by clicking on all pictures with a traffic light, you are helping an algorithm fine-tune its ability to recognize a traffic light.)

Sam, and every other child, can only learn words from others. By following someone else's gaze, a child figures out what the other is referring to. "Aha, that is what they are talking about!" Eye gazing helps narrow the options, making it easier to associate sounds with objects. A human infant's brain excels at breaking the language code, but it does need a lot of input. Social interactions are key, and the more diverse the interactions, the easier learning becomes. Our brain is a social brain, shaped and maintained by interactions with others.

◆

Who needs a village? We all do. Our young, from an evolutionary perspective, were supposed to benefit from exposure to a wide variety of adults and other children. Those attachments and experiences are quite literally what made us who we are today.

The modern-day buzz words *diversity* and *safe space* describe what is needed in childhood: a diversity in social experiences in a safe environment. I say forget about the current push to start "teaching" children earlier and earlier. Instead, we should take inspiration from the northern European countries where preschool extends up to the age of seven and children are encouraged to play and explore. Play gives them the skills needed for future learning. Kindergarten, done well, is a good alternative to the large extended family from days gone by.

Taking care of human infants is so singularly difficult that evolution had to craft a completely new tool to aid the effort. It's a tool that harnesses the mathematics of infinity to let us say what can't be shown: the future and the hypotheticals involved in planning. This book, and the research it contains, would be impossible without it. *Language* was arguably a prerequisite for everything science and society have achieved.

It started small. Very small. With our underbaked infants.

Acknowledgments

Sometimes buried within my inbox of boring, insignificant, and occasionally downright annoying emails, a gem appears. I received such a gem on November 20, 2016: an email from Raghavendra Gadagkar inviting me to apply for a fellowship to join the Wissenschaftskolleg zu Berlin. The Wissenschaftskolleg is better known as WIKO, and Professor Gadagkar is one of its permanent fellows. I'd heard of WIKO, an academic's version of paradise: an institute whose raison d'être is to allow academics (and artists) to spend ten months working on whatever takes their fancy, unencumbered by meetings, commuting, and other distractions. Who wouldn't want to join! And so it happened that in August 2020 (things take time), my husband and fellow academic Ben Oldroyd and I moved from Australia to Berlin to take up our fellowships. We were lucky that we both were asked to apply to WIKO and, even better, both succeeded. Having written too many academic papers read by the handful of colleagues interested in the same niche topic, I wanted to dip my toes into writing a book that anyone interested could read. The result of that ambition you now hold in your hands.

WIKO gave me the mental space and the intellectual environment to embark on book writing. An immensely important part of that intellectual environment was our Not Your Average Book Club, a self-selected bunch of fellows, Ben Oldroyd,

Shamil Jeppie, Anna Frebel, and Hakan Ceylan, all of whom dreamed of writing a popular book one day. We probably would not have gotten anywhere if it weren't for Daniel Schönpflug's excellent mentorship. Daniel is a celebrated historian, writer of popular books and film scripts. Daniel instilled in us the idea that for every page one should generate a scene in the reader's mind—just like a scene in a film. Thank you, Daniel, for your mentorship and patience. A very special thank you is also due to the library staff at WIKO, who managed to get me every single book or article I wanted within no time.

Writing a book is one thing. Getting it published is another. This book would never have seen the light of day if it weren't for a conversation with philosopher of biology Peter Godfrey-Smith. Peter mentioned that he had a list somewhere of literary agents who take on sciencey people. When I had run out of agency options and had recovered from the rejections, I dropped Peter a line, reminding him of that list he had promised to send me. He never found the list, but he did give me two names. "The first guy is new to the profession," he wrote, "and a philosopher too." *Great*, I thought. *A newbie might have the time and mental space to consider my proposal.* Well, the newbie wasn't as captured by my proposal as I had hoped, but apparently he saw a kernel he could work with. And so Alexis (Alexi) Burgess of Park & Fine Literary and Media became my agent. Thanks to his probing questions was I able to home in on what became the book's backbone. I do not think I would have successfully revealed that backbone without Alexi. Alexi's ability to distill a logical argument from my many ramblings proved phenomenal.

When Alexi left Park & Fine Literary and Media, Abigail

(Abby) Koons and Ben Kaslow-Zieve became my indispensable guides through publishingland. A writer, especially a novice, needs people like Abby and Ben for whom a stupid question does not seem to exist (trust me, I tried). Abby in particular has been my rock through some turbulent times, always making the time and effort to check in with me.

Every book needs an editor who believes in the project. For this book that editor is Stephen Morrow. Stephen believed in the book from day one, and his trust strengthened my own belief in the project. I am very lucky that Stephen wanted to take the book with him when he moved from Dutton (Penguin Random House) to Simon & Schuster. His insightful suggestions, often made with a good dose of humor, made the book a much better read. Jennie Miller, too, made many valuable suggestions. Her talent is the ability to see how small changes can make a huge difference. Jennie then took the book under her wing and made it fly. Finally, Carole Berglie's excellent copyediting ironed out all the remaining niggles. If some remain, they are entirely my doing.

I was lucky to have Willemijn Calis and Boris Yagound willing to work with me on the illustrations. Being a biologist himself, Boris was able to translate my rough and rather vague idea into a wonderful picture. Make sure you find Boris if you need a biology-type illustration. My brother, Rodger Beekman, came up with the idea for the cover illustration.

Three people read through earlier versions of the manuscript, sometimes more than once. Their critical eyes helped me iron out some ambiguities and improved the structure of some chapters. A huge thank you to Ben Oldroyd, Ros Gloag, and Evan McGregor. Ben also saved me from potential embarrassment by pointing out some woolly thinking especially concerning molecular biology and genetics.

Writing is a lonely pursuit. I am blessed with a partner who knows that a book doesn't write itself. He's been there. While I tried to craft sentences into paragraphs, paragraphs into chapters, and chapters into this book, Ben took care of everything else. And trust me, there is a lot that needs taking care of around our place.

Speaking of care. As this book argues, children need a lot of it. My daughters Janneke and Willemijn would not have become the beautiful young women they are without the love and commitment from Johan Calis and Pam van Stratum. Sadly my mother, Yvonne Beekman, grandmother extraordinaire, is no longer with us to witness the blossoming of the granddaughters she helped raise.

APPROXIMATE TIMING OF KEY CHANGES THAT

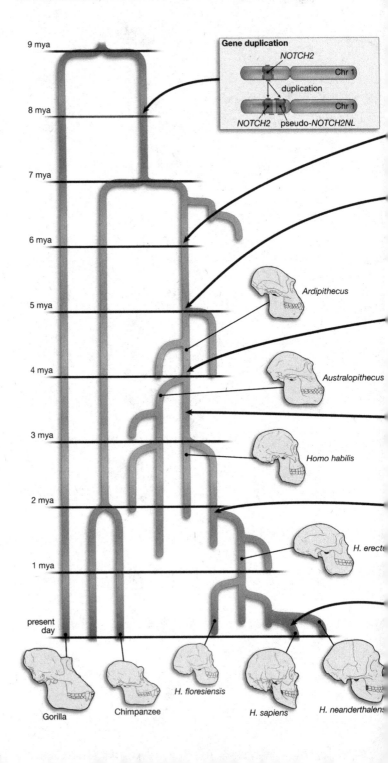

LED US TO BECOME WHAT WE ARE TODAY

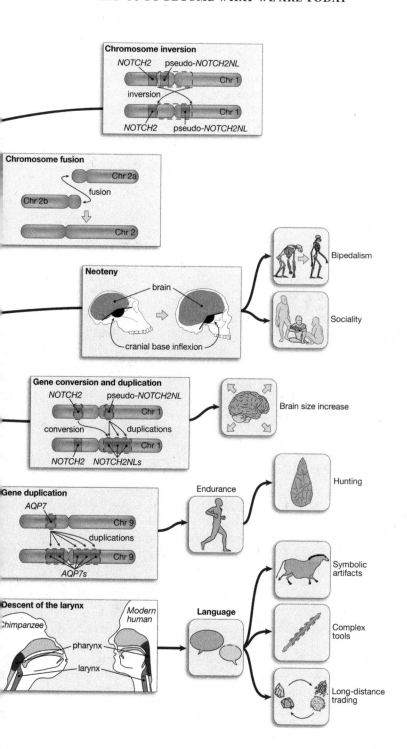

Notes

INTRODUCTION: SEE HOW IT BEGINS

2 **understand what another person is thinking:** H. Clark Barrett, Tanya Broesch, Rose M. Scott, Zijing He, Renée Baillargeon, Di Wu, et al., "Early False-Belief Understanding in Traditional Non-Western Societies," *Proceedings of the Royal Society B: Biological Scisciences* 280, no. 1755 (2013): 20122654.

5 **as a student of von Frisch discovered:** A student of von Frisch's, Martin Lindauer, was intrigued by seeing dancing bees covered in soot, as if they had been sweeping chimneys. What food could be found in a smokestack? Lindauer needed to understand. He set up a swarm of bees on the grounds of a little village called Feld-Moching, then just outside Munich. Lindauer didn't dress in lederhosen, since he was German rather than Austrian, but he used to wear a white lab coat in the field. As his swarm took off, he shed his shoes and ran after them, unaware of the presence of a nearby psychiatric asylum. The next day the local paper ran a report that one of the patients had escaped. What if Lindauer had been caught by the police as he ran barefoot after something he alone could see? "No, Sir, I am not mentally ill but, rather, a scientist following a swarm of bees to see if their dance tells the truth." Yeah, right. . . .

6 **dance language:** Madeleine Beekman and Benjamin P. Oldroyd, "Different Bees, Different Needs: How Nest-Site Requirements Have Shaped the Decision-Making Processes in Homeless Honeybees (*Apis* spp.)," *Philosophical Transactions of the Royal Society B: Biological Sciences* 373, no. 1746 (2018): 20170010.

6 **research on ants:** Madeleine Beekman and Audrey Dussutour, "How to Tell Your Mates—Costs and Benefits of Different Recruitment Mechanisms," in *Food Exploitation by Social Insects: An*

Ecological, Behavioral, and Theoretical Approach, ed. S. Jarau and M. Hrncir (CRC Press, 2009), 115–34.

7 **small genetic change:** Boris Yagound, Kathleen A. Dogantzis, Amro Zayed, Julianne Lim, Paul Broekhuyse, Emily J. Remnant, et al., "A Single Gene Causes Thelytokous Parthenogenesis, the Defining Feature of the Cape Honeybee *Apis mellifera capensis*," *Current Biology* 30, no. 12 (2020): 2248–2259.e6.

ONE: THE 1 PERCENT

13 **Nobel Laureate Niko Tinbergen:** Tinbergen was one of three recipients of the 1973 Nobel Prize in Physiology and Medicine. The others were Karl von Frisch (mentioned in the Introduction) and Konrad Lorenz.

13 **gene's-eye view of evolution:** Richard Dawkins, *An Appetite for Wonder: The Making of a Scientist* (HarperCollins, 2013).

16 **Darwin's *On the Origin of Species*:** Charles R. Darwin, *On the Origin of Species by Means of Natural Selection* (John Murray, 1859).

16 **Dawkins's *The Selfish Gene*:** Richard Dawkins, *The Selfish Gene* (Oxford University Press, 1976).

16 **"nature's complete genetic blueprint":** "Human Genome Project Timeline," National Human Genome Research Institute, July 5, 2022, https://www.genome.gov/human-genome-project/timeline.

16 **a "fortune-telling" device:** Robert Plomin, *Blueprint: How DNA Makes Us Who We Are* (Allen Lane, 2018).

17 **Wellcome Trust Case Control Consortium:** The Wellcome Trust Case Control Consortium, "Genome-Wide Association Study of 14,000 Cases of Seven Common Diseases and 3,000 Shared Controls," *Nature* 447, no. 7145 (2007): 661–78.

18 **"and costs only £100":** Plomin, *Blueprint*, vii.

19 **UK Biobank data set:** This is a large-scale biomedical database and research resource containing in-depth genetic and health information collected from half a million UK participants; see https://www.ukbiobank.ac.uk/.

19 **120 loci associated with income:** W. David Hill, Neil M. Davies, Stuart J. Ritchie, Nathan G. Skene, Julien Bryois, Steven Bell, et al., "Genome-Wide Analysis Identifies Molecular Systems and

149 Genetic Loci Associated with Income," *Nature Communications* 10, no. 1 (2019): 5741.
21 **Mary-Claire King and Allan Wilson:** Mary-Claire King and Allan Charles Wilson, "Evolution at Two Levels in Humans and Chimpanzees," *Science* 188, no. 4184 (1975): 107–16.
24 **group of scientists compared:** Derek E. Wildman, Monica Uddin, Guozhen Liu, Lawrence I. Grossman, and Morris Goodman, "Implications of Natural Selection in Shaping 99.4% Nonsynonymous DNA Identity Between Humans and Chimpanzees: Enlarging Genus *Homo*," *Proceedings of the National Academy of Sciences* 100, no. 12 (2003): 7181–88.
26 **skeletons of a gibbon:** Thomas H. Huxley, *Evidence as to Man's Place in Nature* (J.M. Dent & Sons, 1927), xiv.
26 **"existing ape or monkey":** Charles Darwin, *The Descent of Man, and Selection in Relation to Sex* (John Murray, 1871), 200.
26 **Linnaeus systematically grouped organisms:** Gordon McGregor Reid, "Carolus Linnaeus (1707-1778): His Life, Philosophy and Science and Its Relationship to Modern Biology and Medicine," *Taxon* 58, no. 1 (2009): 18–31.
27 **laughably inaccurate:** Huxley, *Evidence as to Man's Place*, 12.
29 **"The difference in mind":** Darwin, *Descent of Man*, 106.
29 **"man's own instinctive cries":** Darwin, *Descent of Man*, 57.
29 **"strengthened and perfected":** Darwin, *Descent of Man*, 58.
30 **Marie Eugène François Thomas Dubois:** Pat Shipman, *The Man Who Found the Missing Link: Eugène Dubois and His Lifelong Quest to Prove Darwin Right* (Harvard University Press, 2002).
30 ***History of Creation*:** Ernst H. Haeckel, *The History of Creation, or the Development of the Earth and Its Inhabitants by the Action of Natural Causes* (Henry S. King, 1876), vols. 1 and 2.
31 **"one of the chief human characteristics":** Ernst Haeckel, *The Evolution of Man* (Watts & Co., 1912), https://www.gutenberg.org/files/8700/8700-h/8700-h.htm.
32 **two Neanderthal skeletons:** Julien Fraipont and Max Lohest, *La race humaine de Néanderthal ou de Canstadt en Belgique: recherches ethnographiques sur des ossements humains découverts dans les dépôts quaternaires d'une grotte à Spy et détermination de leur âge géologique* (Vanderpoorten, 1887).

32 **"save in a dream":** For more on Rudolf Virchow, see https://geniuses.club/genius/rudolf-virchow.

33 **a cranium, a tooth, and a femur:** The originals are kept in a museum in Leiden, the Netherlands. See Naturalis Biodiversity Center, https://www.naturalis.nl/en.

34 **three species of *Homo*:** Ernst Mayr, "Taxonomic Categories in Fossil Hominids," in *Origin and Evolution of Man* (Cold Spring Harbor Symposia on Quantitative Biology, 1950): 15:109–18.

35 **They called it *Homo floresiensis*:** P. Brown, T. Sutikna, M. J. Morwood, R. R. Soejono, E. Wayhu Saptomo, et al., "A New Small-Bodied Hominin from the Late Pleistocene of Flores, Indonesia," *Nature* 431, no. 7012 (2004): 1055–61.

36 **artifacts and animal remains:** Debbie Argue, Denise Donlon, Colin Groves, and Richard Wright, "*Homo floresiensis*: Microcephalic, Pygmoid, *Australopithecus*, or *Homo*?," *Journal of Human Evolution* 51, no. 4 (2006): 360–74.

37 **deformed modern human:** T. Jacob, E. Indriati, R. P. Soejono, K. Hsü, D. W. Frayer, R. B. Eckhardt, et al., "Pygmoid Australomelanesian *Homo sapiens* Skeletal Remains from Liang Bua, Flores: Population Affinities and Pathological Abnormalities," *Proceedings of the National Academy of Sciences* 103, no. 36 (2006): 13421–26.

37 **a new species:** M. J. Morwood, R. P. Soejono, R. G. Roberts, T. Sukikna, C. S. M. Burney, K. E. Westaway, et al., "Archaeology and Age of a New Hominin from Flores in Eastern Indonesia," *Nature* 431, no. 7012 (2004): 1087–91.

38 **the mystery of human evolution:** Svante Pääbo, *Neanderthal Man: In Search of Lost Genomes* (Basic Books, 2014).

38 **when he read the publication:** S. Pääbo (1985). "Molecular Cloning of Ancient Egyptian Mummy DNA," *Nature* 314 (1985): 644–45.

39 **one million years old:** Tom van der Valk, Patricia Pecnerova, David Diez-del-Molino, Anders Bergstrom, Jonas Oppenheimer, Stefanie Hartmann et al., "Million-Year-Old DNA Sheds Light on the Genomic History of Mammoths," *Nature* 591, no. 7849 (2021): 265–69.

40 **Reich did find in nuclear DNA:** David Reich, *Who We Are*

and *How We Got Here: Ancient DNA and the New Science of the Human Past* (Pantheon, 2018).

42 **Falconer's visit:** Jeremy M. DeSilva, Preface, in *A Most Interesting Problem: What Darwin's Descent of Man Got Right and Wrong About Evolution*, ed. Jeremy M. DeSilva (Princeton Universtiy Press, 2021), xiii–xxi.

TWO: OUR ORIGINAL CHILDCARE PROBLEM

46 **mimicking the shape of the leaves:** Benji Jones, "The Mystery of the Mimic Plant," *Vox*, January 11, 2023, https://www.vox.com/down-to-earth/2022/11/30/23473062/plant-mimicry-boquila-trifoliolata.

47 **Ardi's discovery changed everything:** Yohannes Haile-Selassie, "Charles Darwin and the Fossil Evidence for Human Evolution," in *A Most Interesting Problem: What Darwin's Descent of Man Got Right and Wrong About Evolution*, ed. J. M. DeSilva (Princeton Universtiy Press, 2021), 82–102.

49 **authenticity of Piltdown Man:** Jessie Szalay, "Piltdown Man: Infamous Fake Fossil," *LiveScience*, September 20, 2016, https://www.livescience.com/56327-piltdown-man-hoax.html.

49 **the fossils were fakes:** "Science: End as a Man," *Time*, November 30, 1953, https://content.time.com/time/subscriber/article/0,33009,823171,00.html.

50 **early human ancestor species:** Haile-Selassie, "Charles Darwin and the Fossil Evidence for Human Evolution."

52 **muscles that extend limbs:** Laura Tobias Gruss and Daniel Schmitt, "The Evolution of the Human Pelvis: Changing Adaptations to Bipedalism, Obstetrics and Thermoregulation," *Philosophical Transactions of the Royal Society B: Biological Sciences* 370, no. 1663 (2015): 20140063.

54 **fit their shoulders through it:** Jeremy M. DeSilva, Natalie M. Laudicia, Karen R. Rosenberg, and Wenda R. Trevathan, "Neonatal Shoulder Width Suggests a Semirotational, Oblique Birth Mechanism in *Australopithecus afarensis*," *Anatomical Record* 300, no. 5 (2017): 890–99.

55 **had lost most of her body hair:** We suspect Lucy was more or less

hairless because of the genetics of two species of lice—head lice and pubic lice. Head lice are closely related to the lice found on chimpanzees (and the body or clothes lice are related to head lice and emerged when, you guessed, humans started to wear clothes), but pubic lice are more closely related to gorilla lice. Most likely, the head lice "retreated" to the head once Lucy and her kin lost their body hair, and the pubic area became an open niche for gorilla lice. How that host-jump happened is anyone's guess—or nightmare; see David L. Reed, Jessica E. Light, Julie M. Allen, and Jeremy J. Kirchman, "Pair of Lice Lost or Parasites Regained: The Evolutionary History of Anthropoid Primate Lice," *BMC Biology* 5, no. 1 (2007): 5–7.

55 **incompatible with walking upright:** Lia Q. Amaral, "Mechanical Analysis of Infant Carrying in Hominoids," *Naturwissenschaften* 95, no. 4 (2008): 281–92.

56 **nothing compared with our brain:** Kate M. Lesciotto and Joan T. Richtsmeier, "Craniofacial Skeletal Response to Encephalization: How Do We Know What We Think We Know?," *American Journal of Physical Anthropology* 168, no. S67 (2019): 27–46.

57 **matches that of chimpanzees:** M. Boeckle, M. Schiestl, A. Frohnwieser, R. Gruber, R. Miller, T. Suddendorf, et al., "New Caledonian Crows Plan for Specific Future Tool Use," *Proceedings of the Royal Society B: Biological Sciences* 287, no. 1938 (2020): 20201490.

57 **back to the nest:** Gábor Módra, I. Maák, Á. Lőrincz, and Gábor Lőrinczi, "Comparison of Foraging Tool Use in Two Species of Myrmicine Ants (Hymenoptera: Formicidae)," *Insectes Sociaux* 69, no. 1 (2022): 5–12.

57 **the nest entrance of neighbors:** Enrico Schifani, Cristina Castracani, Daniele Giannetti, Fiorenza A. Spotti, Alessandra Mori, and Donato A. Grasso, "Tool Use in Pavement Battles Between Ants: First Report of *Tetramorium immigrans* (Hymenoptera, Formicidae) Using Soil-Dropping as an Interference Strategy," *Insectes Sociaux* 69, no. 4 (2022): 355–59.

58 **"free use of the arms and hands":** Charles Darwin, *The Descent of Man, and Selection in Relation to Sex* (John Murray, 1871), 145.

60 ***Homo erectus* was the first one:** Dennis M. Bramble and Daniel

E. Lieberman, "Endurance Running and the Evolution of *Homo*," *Nature* 432, no. 7015 (2004): 345–52.

63 **just to maintain their brain:** Daniel E. Lieberman, *The Evolution of the Human Head* (Belknap Press of Harvard University Press, 2011).

63 **the same weight:** Jeremy M. DeSilva, "A Shift Toward Birthing Relatively Large Infants Early in Human Evolution," *Proceedings of the National Academy of Sciences* 108, no. 3 (2011): 1022–27.

63 **exceptionally fat:** Christopher W. Kuzawa, "Adipose Tissue in Human Infancy and Childhood: An Evolutionary Perspective," *American Journal of Physical Anthropology*, Suppl. 27 (1998): 177–209.

63 **"Human Babies as Embryos":** Stephen J. Gould, *Ever Since Darwin: Reflections on Natural History* (W.W. Norton, 1977), chap. 8.

64 **Gould but by others:** Holly M. Dunsworth, "There Is No Evolutionary 'Obstetrical Dilemma,'" in *The Routledge Handbook of Anthropology and Reproduction*, ed. C. Tomori and S. Han (Taylor & Francis, 2021), chap. 27.

64 **a little silly:** Dunsworth, "There Is No Evolutionary 'Obstetrical Dilemma.'"

65 **future adult size:** Jeremy M. DeSilva and Julie Lesnik, "Chimpanzee Neonatal Brain Size: Implications for Brain Growth in *Homo erectus*," *Journal of Human Evolution* 51, no. 2 (2006): 207–12.

65 **28 percent of its adult size:** DeSilva and Lesnik, "Chimpanzee Neonatal Brain Size."

65 **first eighteen months:** Robert A. Foley and P. Clive Lee, "Ecology and Energetics of Encephalization in Hominid Evolution," *Philosophical Transactions of the Royal Society B: Biological Sciences* 334, no. 1270 (1991): 223–31.

66 **thirty-seven days of pregnancy:** Holly M. Dunsworth, Anna G. Warrener, Terrence Deacon, Peter T. Ellison, and Herman Pontzer, "Metabolic Hypothesis for Human Altriciality," *Proceedings of the National Academy of Sciences* 109, no. 38 (2012): 15212–16.

66 **energetics of gestation and growth (EGG):** Dunsworth et al., "Metabolic Hypothesis for Human Altriciality."

67 **inter-birth interval:** Louise Humphrey, "Weaning Behavior in

Human Evolution," *Seminars in Cell & Developmental Biology* 21, no. 4 (2010): 453–61.
69 **"with strictly social animals"**: Darwin, *Descent of Man*, 156.
70 **"his higher mental powers"**: Darwin, *Descent of Man*, 146.

THREE: BEAT OF A DIFFERENT DRUM

72 **an important process in evolution:** Stephen J. Gould, "A Biological Homage to Mickey Mouse," *Ecotone* 4 (2008): 333–40. The essay first appeared in *Natural History*, May 1979, as "This View of Life."
73 **the bizarre cycle again:** Ibrahim A. Ibrahim and Amin M. Gad, "The Occurrence of Paedogenesis in *Eristalis* Larvae (Diptera: Syrphidae)," *Journal of Medical Entomology* 12, no. 2 (1975): 268–68.
73 **"temporal retardation of development":** Stephen J. Gould, *Ontogeny and Phylogeny* (Belknap Press of Harvard University Press, 1977), 365.
74 ***Sahelanthropus tchadensis*:** See G. A. Russo and E. C. Kirk, "Foramen magnum Position in Bipedal Mammals," *Journal of Human Evolution* 65, no. 5 (2013): 656–70.
74 **A video of Pedals:** See "Pedals Bipedal Bear Sighting," posted June 22, 2016, by NJ.com, YouTube, https://www.youtube.com/watch?v=Mk-HHyGRSRw.
76 **"mischievous in its tendency":** Louis Agassiz, "Prof. Agassiz on the Origin of Species," *American Journal of Science and Arts* 30, no. 2 (1860); also see Edward Lurie, "Louis Agassiz and the Idea of Evolution," *Victorian Studies* 3, no. 1 (1959): 87–108.
76 **Darwin used this anecdote:** Darwin included this anecdote in the sixth edition of his *On the Origin of Species,* 1872, page 419.
76 **a four-volume book:** Beautifully described in Rebecca Stott, *Darwin and the Barnacle: The Story of One Tiny Creature and History's Most Spactacular Scientific Breakthrough* (W. W. Norton, 2004).
77 **a special group:** Hsiu-Chin Lin, Gregory A. Kobasov, and Benny K. K. Chan, "Phylogenetic Relationships of Darwin's 'Mr. Arthrobalanus': The Burrowing Barnacles (Cirripedia: Acrothoracica)," *Molecular Phylogenetics and Evolution* 100 (July 2016): 292–302.

78 **got him into trouble:** Nick Hopwood, "Pictures of Evolution and Charges of Fraud: Ernst Haeckel's Embryological Illustrations," *Isis* 97 (June 2006): 260–301.

78 **access to a human embryo:** Ernst Mayr, *What Evolution Is* (Basic Books, 2001).

79 **all organisms except bacteria and viruses:** It seems that many plants and some groups of fungi have lost their equivalent of Hedgehog genes; see Thomas R. Bürglin, "The Hedgehog Protein Family," *Genome Biology* 9, no. 11 (2008): 241.

80 **the *Hox* gene family:** Rushikesh Sheth, Damien Grégorie, Annie Dumouchel, Martina Scotti, Jessica My Trang Pham, Stephen Nemec, et al., "Decoupling the Function of Hox and Shh in Developing Limb Reveals Multiple Inputs of Hox Genes on Limb Growth," *Development* 140, no. 10 (2013): 2130–38.

82 ***Hes7* oscillations differ among taxa:** M. Marshall, "These Cellular Clocks Help Explain Why Elephants Are Bigger than Mice," *Nature* 592 (April 27, 2021): 682–84.

82 **"What can be more curious":** Charles R. Darwin, *On the Origin of Species by Means of Natural Selection* (John Murray, 1859), 517.

83 **similar regulatory networks:** Neil Shubin, Cliff Tabin, and Sean Carroll, "Deep Homology and the Origins of Evolutionary Novelty," *Nature* 457, no. 7231 (2009): 818–23.

83 **"indelible stamp of his lowly origin":** Charles Darwin, *The Descent of Man, and Selection in Relation to Sex* (John Murray, 1871), 406.

84 **that are quadrupedal:** G. A. Russo and E. C. Kirk, "Foramen Magnum Position in Bipedal Mammals," *Journal of Human Evolution* 65, no. 5 (2013): 656–70.

85 **normal development:** David M. Kingsley, "What Do BMPs Do in Mammals? Clues from the Mouse Short-Ear Mutation," *Trends in Genetics* 10, no. 1 (1994): 16–21.

87 **penises are spineless:** Cory Y. McLean, Philip L. Reno, Alex A. Pollen, Abraham I. Bassan, Terence D. Capellini, Catherine Guenther, et al., "Human-Specific Loss of Regulatory DNA and the Evolution of Human-Specific Traits," *Nature* 471, no. 7337 (2011): 216–19.

88 **into a human-like foot:** Vahan B. Indjeian, Garrett A. Kingman, Felicity C. Jones, Catherine A. Guenther, Jan Grimwood, Jeremy Schmutz, et al., "Evolving New Skeletal Traits by cis-Regulatory Changes in Bone Morphogenetic Proteins," *Cell* 164, no. 1 (2016): 45–56.

89 **wrists, and shoulders:** Shyam Prabhakar, Axel Visel, Jennifer A. Akiyama, Malak Shoukry, Keith D. Lewis, Amy Holt, et al., "Human-Specific Gain of Function in a Developmental Enhancer," *Science* 321, no. 5894 (2008): 1346–50.

89 **expression in embryonic tissue:** The technique used is described in R. Kothary, S, Clapoff, S. Darling, M. D. Perry, L. A. Moran, and J. Rossant, "Inducible Expression of an hsp68-lacZ Hybrid Gene in Transgenic Mice," *Development* 105, no. 4 (1989): 707–14.

90 **humans have only 23:** J. J. Yunis and O. Prakash, "The Origin of Man: A Chromosomal Pictorial Legacy," *Science* 215, no. 4539 (1982): 1525–30; Yuxin Fan, Elena Linardopoulou, Cynthia Friedman, Eleanor Williias, and Barbara J. Trask, "Genomic Structure and Evolution of the Ancestral Chromosome Fusion Site in 2q13–2q14.1 and Paralogous Regions on Other Human Chromosomes," *Genome research* 12, no. 11 (2002): 1651–62.

90 **lead to speciation events:** Anne-Marie Dion-Côté and Daniel A. Barbash, "Beyond Speciation Genes: An Overview of Genome Stability in Evolution and Speciation," *Current Opinion in Genetics & Development* 47 (December 2017): 17–23.

93 **3 billion swipes globally:** Dougal Shaw, "Coronavirus: Tinder Boss Says 'Dramatic' Changes to Dating," *BBC News*, May 20, 2020, https://www.bbc.com/news/business-52743454.

93 **above-average desirability scores:** Austin Carr, "I Found Out My Tinder Rating and Now I Wish I Hadn't," *Fast Company*, November 1, 2016, www.fastcompany.com/3054871/whats-your-tinder-score-inside-the-apps-internal-ranking-system.

96 **become even better dispersers:** Richard Shine, Gregory Brown, Benjamin Phillips, and David Wake, "An Evolutionary Process that Assembles Phenotypes Through Space Rather than Through Time," *Proceedings of the National Academy of Sciences* 108, no. 14 (2011): 5708–11.

97   **speed up the process:** G. L. Bush, S. M. Case, A. C. Wilson, and J. L. Patton, "Rapid Speciation and Chromosomal Evolution in Mammals," *Proceedings of the National Academy of Sciences* 74, no. 9 (1977): 3942–46.

100 **humans are found everywhere:** Rory Bowden, Tammie MacFie, Simon Myers, Garrett Hellenthal, Eric Nerrienet, Ronald Bontrop, et al., "Genomic Tools for Evolution and Conservation in the Chimpanzee: *Pan troglodytes* ellioti Is a Genetically Distinct Population," *PLoS Genetics* 8, no. 3 (2012): e1002504.

FOUR: MIND BLOWN

101 **"intelligent" behaviors:** Madeleine Beekman and Tanya Latty, "Brainless but Multi-Headed: Decision-Making by the Acellular Slime Mould *Physarum polycephalum*," *Journal of Molecular Biology* 427, no. 23 (2015): 3734–43; Jules Smith-Ferguson and Madeleine Beekman, "Who Needs a Brain? Slime Moulds, Behavioural Ecology and Minimal Cognition," *Adaptive Behavior* 28, no. 6 (2019): 1059712319826537.

105 **members of the genus on record:** Daniel E. Lieberman, "Speculations about the Selective Basis for Modern Human Craniofacial Form," *Evolutionary Anthropology: Issues, News, and Reviews* 17, no. 1 (2008): 55–68.

105 **"rationals, but as animals":** Thomas Paine, *The Rights of Man* (J. S. Jordan, 1791), 84.

108 **which I guess in essence it is:** Edward B. Lewis, "Alfred Henry Sturtevant, November 21, 1891–April 5, 1970, https://nap.natio nalacademies.org/read/9650/chapter/20#351, reprint from *Dictionary of Scientific Biography* (Scribner's, 1976), 13:133–38.

110 **chromosome structure around the gene:** Mariana F. Wolfner and Danny E. Miller, "Alfred Sturtevant Walks into a Bar: Gene Dosage, Gene Position, and Unequal Crossing over in *Drosophila*," *Genetics* 204, no. 3 (2016): 833–35.

110 **in Sturtevant's words:** A. H. Sturtevant, "The Effects of Unequal Crossing over at the Bar Locus in Drosophila," *Genetics* 10, no. 2 (1925): 117–47, https://doi.org/10.1093/genetics/10.2.117.

111 **from continuing her work:** Sandeep Ravindran, "Barbara McClintock and the Discovery of Jumping Genes," *Proceedings of the National Academy of Sciences* 109, no. 50 (2012): 20198–99.

111 **rife with transposable elements:** Roy J. Britten, "Transposable Element Insertions Have Strongly Affected Human Evolution," *Proceedings of the National Academy of Sciences USA* 107, no. 46 (2010): 19945–48.

111 **glued back on upside down:** Justyna M. Szamalek, Violaine Goidts, David N. Cooper, Horst Hameister, and Hildegard Kehjrer-Sawatzki, "Characterization of the Human Lineage-Specific Pericentric Inversion that Distinguishes Human Chromosome 1 from the Homologous Chromosomes of the Great Apes," *Human Genetics* 120, no, 1 (2006): 126–38.

112 **still an undergraduate student:** Mark Kirkpatrick, "How and Why Chromosome Inversions Evolve," *PLoS Biology* 8, no. 9 (2010): e1000501.

112 **"gene conversion":** David Benovoy and Guy Drouin, "Ectopic Gene Conversions in the Human Genome," *Genomics* 93, no. 1 (2009): 27–32.

112 **disproportionate number of genes:** Majesta O'Bleness, Veronica B. Searles, Ajit Varki, Pascal Gagneux, and James M. Sikela, "Evolution of Genetic and Genomic Features Unique to the Human Lineage," *Nature Reviews Genetics* 13, no. 12 (2012): 853–66.

114 **the growth of the brain:** Ian T. Fiddes, Gerrald A. Lodewijk, Meghan Mooring, Colleen M. Bosworth, Adam D. Ewing, Gary L. Mantalas, et al., "Human-Specific *NOTCH2NL* Genes Affect Notch Signaling and Cortical Neurogenesis," *Cell* 173, no. 6 (2018): 1356–69.e22.

115 **three functional versions of *NOTCH2NL*:** Fiddes et al., "Human-Specific *NOTCH2NL* Genes."

116 **look at very young embryos:** Silvia Benito-Kwiecinski, Stefano Giandomenico, Magdalena Sutcliffe, Erlend Riis Paula Freire-Pritchett, Iva Kelava, et al., "An Early Cell Shape Transition Drives Evolutionary Expansion of the Human Forebrain," *Cell* 184, no. 8 (2021): 2084–2102.e19.

118 **than any other animal:** Yana G. Kamberov, Samantha M. Guhan, Alessandra DeMarchis, Judy Jiang, Sara S. Wright, Bruce A.

Morgan, et al., "Comparative Evidence for the Independent Evolution of Hair and Sweat Gland Traits in Primates," *Journal of Human Evolution* 125 (December 2018): 99–105.

118 **increase its brain size:** Daniel E. Lieberman, *The Evolution of the Human Head* (Belknap Press of Harvard University Press, 2011).

119 **rife with duplicated genes:** Marianna Paulis, Mirella Bensi, Daniela Moralli, Luigi de Carli, and Elena Raimondi, "A Set of Duplicons on Human Chromosome 9 Is Involved in the Origin of a Supernumerary Marker Chromosome," *Genomics* 87, no. 6 (2006): 747–57.

119 **conversions and duplications:** Laura Dumas, Young Kim, Anis Karimpour-Fard, Michael Cox, Janet Hopkins, Jonathan Pollack, et al., "Gene Copy Number Variation Spanning 60 Million Years of Human and Primate Evolution," *Genome Research* 17, no. 9 (2007): 1266–77.

119 **modern-day humans:** Simon Neubauer, Jean-Jacques Hublin, and Philipp Guz, "The Evolution of Modern Human Brain Shape," *Science Advances* 4, no. 1 (2018): eaao5961.

120 **a contradiction in terms:** Lieberman, *Evolution of the Human Head*.

121 **grow at different rates:** Xavier Penin, Christine Berge, and Michel Baylac, "Ontogenetic Study of the Skull in Modern Humans and the Common Chimpanzees: Neotenic Hypothesis Reconsidered with a Tridimensional Procrustes Analysis," *American Journal of Physical Anthropology* 118, no. 1 (2002): 50–62.

122 **Neus Martínez-Abadías and her colleagues:** Neus Martínez-Abadías, Mireia Esparza, Torstein Sjofold, Rolando Gonzalez-José, Mauro Santos, Miguel Hernandez, et al., "Pervasive Genetic Integration Directs the Evolution of Human Skull Shape," *Evolution* 66, no. 4 (2012): 1010–23.

126 **the skull's interconnectedness:** Daniel E. Lieberman, Benedikt Haligrimsson, Wei Liu, Trish E. Parsons, and Heather A. Jamniczky, "Spatial Packing, Cranial Base Angulation, and Craniofacial Shape Variation in the Mammalian Skull: Testing a New Model Using Mice," *Journal of Anatomy* 212, no. 6 (2008): 720–35.

127 **different parts of the embryo:** Mariya P. Dobreva, Jasmin Camacho, and Arkhat Abzhanov, "Time to Synchronize Our Clocks:

Connecting Developmental Mechanisms and Evolutionary Consequences of Heterochrony," *Journal of Experimental Zoology Part B: Molecular and Developmental Evolution* 338, nos. 1–2 (2022): 87–106.

128 **pack in a large brain:** Daniel E. Lieberman, "Speculations About the Selective Basis for Modern Human Craniofacial Form," *Evolutionary Anthropology: Issues, News, and Reviews* 17, no. 1 (2008): 55–68.

FIVE: LOUD MOMS

135 **evolutionary biologists decided to test:** Michael Griesser, Szymon Drobniak, Sereina Graber, and Carel van Schaik, "Parental Provisioning Drives Brain Size in Birds," *Proceedings of the National Academy of Sciences* 120, no. 2 (2023): e2121467120.

138 **with a thought experiment:** Sarah B. Hrdy, *Mothers and Others: The Evolutionary Origins of Mutual Understanding* (Belknap Press of Harvard University Press, 2009), 121.

139 **during a dry spell:** Anne E. Pusey, "Warlike Chimpanzees and Peacemaking Bonobos," *Proceedings of the National Academy of Sciences* 119, no. 31 (2022): e2208865119.

140 **different from those of the men:** Richard Borshay Lee and Irven DeVore, eds., *Man the Hunter* (Routledge, 1968).

141 **a recent study:** Ingela Alger, Slimane Dridi, Jonathjan Stieglitz, and Michael Wilson, "The Evolution of Early Hominin Food Production and Sharing," *Proceedings of the National Academy of Sciences* 120, no. 25 (2023): e2218096120.

142 **turns out to be illuminating:** Madeleine Beekman, Michael Thompson, and Marko Jusup, "Thermodynamic Constraints and the Evolution of Parental Provisioning in Vertebrates," *Behavioral Ecology* 30, no. 3 (2019): 583–91.

142 **produce milk for the mini-caecillians:** Pedro L. Mailho-Fontana, Marta Antoniazzi, Guilherme Coelho, Daniel Pimenta, Ligna Fernandes, Alexander Kupfer, et al., "Milk Provisioning in Oviparous Caecilian Amphibians," *Science* 383, no. 6687 (2024): 1092–95.

147 **10 percent of all species:** Hannah E. R. West and Isabella Capellini, "Male Care and Life History Traits in Mammals," *Nature Communications* 7, no. 1 (2016): 11854.

149 **having been evicted:** K. J. MacLeod, J. F. Nielse, and T. H. Clutton-Brock, "Factors Predicting the Frequency, Likelihood and Duration of Allonursing in the Cooperatively Breeding Meerkat," *Animal Behaviour* 86, no. 5 (2013): 1059–67.

151 **lioness nicknamed Kamuniak:** Hrdy, *Mothers and Others*, chap. 7.

153 **a long-term study:** The study is described in Charles A. Nelson, Nathan A. Fox, and Charles H. Zeanah, *Romania's Abandoned Children: Deprivation, Brain Development, and the Struggle for Recovery* (Harvard University Press, 2014).

153 **but never completely:** Charles A. Nelson, Charles Zeanah, Nathan Fox, Petr Marshall, Anna Smyke, and Donald Guthrie, "Cognitive Recovery in Socially Deprived Young Children: The Bucharest Early Intervention Project," *Science* 318, no. 5858 (2007): 1937–40.

154 **childhood is *for* learning:** Alison Gopnik, *The Gardener and the Carpenter: What the New Science of Child Development Tells Us About the Relationship Between Parents and Children* (Farrar, Straus & Giroux, 2016).

154 **implements she needed:** Jennifer Holzhaider, Gavin Hunt, and Russell Gray, "The Development of Pandanus Tool Manufacture in Wild New Caledonian Crows," *Behaviour* 147, nos. 5–6 (2010): 553–86.

158 **we enlisted our parents:** Because human societies rely on grandparents as alloparents, instead of younger helpers, some prefer the term *biocultural reproduction* over *cooperative breeding*. See Barry Bogin, Jared Bragg, and Christopher Kuzawa, "Humans Are Not Cooperative Breeders but Practice Biocultural Reproduction," *Annals of Human Biology* 41, no. 4 (2014): 368–80.

158 **evolved fairly recently:** Rachel Caspari and Sang-Hee Lee, "Older Age Becomes Common Late in Human Evolution," *Proceedings of the National Academy of Sciences* 101, no. 30 (2004): 10895–900.

158 **cultural transmission:** Stanley H. Ambrose, "Paleolithic Technology and Human Evolution," *Science* 291, no. 5509 (2001): 1748–53.
159 **broken her ankle:** Jeremy DeSilva, *First Steps: How Walking Upright Made Us Human* (William Collins, 2022).

SIX: WHO NEEDS HALF A GRAMMAR?

161 **language evolved:** Alfred R. Wallace, *Contributions to the Theory of Natural Selection: A Series of Essays* (Macmillan, 1870).
163 **more than 200 words:** J. Kaminski, J. Call, and J. Fischer, "Word Learning in the Domestic Dog: Evidence for 'fastmapping,'" *Science* 304 (2004): 1682–83, https://doi.org/10.1126/science.1097859.
164 **another 1,022 words:** J. W. Pilley and A. K. Reid, "Border Collie Comprehends Object Names as Verbal Referents," *Behavioural Processes* 86 (2010): 184–95, https://doi.org/10.1016/j.beproc.2010.11.007).
164 **speech and language unit:** J. A. Hurst, M. Baraitser, E. Auger, F. Graham, and S. Norell, "An Extended Family with a Dominantly Inherited Speech Disorder," *Developmental Medicine and Child Neurology* 32, no. 4 (1990): 352–55.
165 **the family's speech disorder:** S. E. Fisher, F. Vargha-Khadem, K. E. Watkins, A. P. Monaco, and M. E. Pembrey, "Localisation of a Gene Implicated in a Severe Speech and Language Disorder," *Nature Genetics* 18, no. 2 (1998): 168–70.
166 ***FOXP2*:** C. S. Lai, S. E. Fisher, J. A. Hurst, F. Vargha-Khadem, and A. P. Monaco, "A Forkhead-Domain Gene Is Mutated in a Severe Speech and Language Disorder," *Nature* 413, no. 6855 (2001): 519–23.
166 **distinctive mutation:** Wolfgang Enard, Molly Przeworski, Simon Fisher, Cecilia Lai, Victor Wiebe, Takashi Kitano, et al., "Molecular Evolution of FOXP2, a Gene Involved in Speech and Language," *Nature* 418, no. 6900 (2002): 869–72.
166 **among others:** Fisher et al., "Localisation of a Gene."
167 **manipulated mice and zebra finches:** W. Enard, "FOXP2 and the Role of Cortico-Basal Ganglia Circuits in Speech and Language Evolution," *Current Opinion in Neurobiology* 21 (2011): 415–24.

167 *FOXP2* **into mice:** Christiane Schreiweis, Ulrich Bornschein, Eric Burguière, Cemil Kerimoglu, Sven Schreiter, Michael Danneann, et al., "Humanized Foxp2 Accelerates Learning by Enhancing Transitions from Declarative to Procedural Performance," *Proceedings of the National Academy of Science USA* 111, no. 39 (2014): 14253–58.

168 **our species:** Elizabeth G. Atkinson, Amanda Audesse, Julia Palacios, Dean Bobo, Ashley Webb, Sohini Ramachandrian, et al., "No Evidence for Recent Selection at FOXP2 Among Diverse Human Populations," *Cell* 174, no. 6 (2018): 1424–35.e15.

169 **all humans share the same mutation:** Atkinson et al., "No Evidence for Recent Selection."

170 **amusing real-life heading:** Steven Pinker, *The Language Instinct: How the Mind Creates Language* (Harper Perennial, 1994), 448.

171 **captive breeding program:** See Commonwealth of Australia 2016 National Recovery Plan for the Regent Honeyeater (*Anthochaera phrygia*), www.agriculture.gov.au/sites/default/files/documents/national-recovery-plan-regent-honeyeater.pdf.

172 **the time of day:** Daniel L. Appleby, Joy Tripovich, Naoli Langmore, Robert Hleinsohn, Benjamin J. Pitcher, and Ross Crates, "Zoo-Bred Female Birds Prefer Songs of Zoo-Bred Males: Implications for Adaptive Management of Reintroduction Programs," *Biological Conservation* 284 (August 2023): 110171.

173 **any human child:** E. Sue Savage-Rumbaugh, J. Murphy, R. A. Sevcik, K. E. Brakke, S. L. Williams, and D. M. Rumbaugh, "Language Comprehension in Ape and Child," *Monographs of the Society for Researchj in Child Development* 58, nos. 3–4 (1993): 1–222.

174 **wonderful videos available:** See, for example, *Kanzi: An Ape of Genius* (NHK Broadcasting, 1993), posted October 19, 2010, in four parts by flactemb, YouTube, https://www.youtube.com/playlist?list=PLB4411A9C9B7CCEEA.

175 **Kanzi's making random choices:** P. T. Schoenemann, "Evidence of Grammatical Knowledge in Apes: An Analysis of Kanzi's Performance on Reversible Sentences," *Frontiers of Psychology* 13 (July 2022): 885605.

176 **facial expression:** Andrew N. Meltzoff and M. Keith Moore,

"Imitation of Facial and Manual Gestures by Human Neonates," *Science* 198, no. 4312 (1977): 75–78.

176 **behave much like a chimpanzee:** These studies are described in Sarah B. Hrdy, *Mothers and Others: The Evolutionary Origins of Mutual Understanding* (Belknap Press of Harvard University Press, 2009).

177 **has been challenged:** Cecilia Heyes, Nick Chater, and Dominic Dwyer, "Sinking In: The Peripheral Baldwinisation of Human Cognition," *Trends in Cognitive Sciences* 24, no. 11 (2020): 893.

178 **intrigued by human faces:** Laura Carnevali, Anna Gui, Emily Jones, and Teresa Farroni, "Face Processing in Early Development: A Systematic Review of Behavioral Studies and Considerations in Times of COVID-19 Pandemic," *Frontiers in Psychology* 13 (February 2022): 778247.

179 **Tomasello's extraordinary experiments:** Michael Tomasello, "A Lecture in Psychology: Origins of Human Cooperation and Morality," *Vimeo*, December 6, 2012, https://vimeo.com/55035079.

181 **the syllable *ba*:** Patricia K. Kuhl, "How Babies Learn Language," *Scientific American*, November 1, 2018, https://www.scientificamerican.com/article/how-babies-learn-language/.

182 **cooperative-eye hypothesis:** Fumihiro Kano, Yuri Kawaguchi, and Yeow Hanling, "Experimental Evidence that Uniformly White Sclera Enhances the Visibility of Eye-Gaze Direction in Humans and Chimpanzees," *eLife* 11 (March 2022): e74086.

183 **In his 1872 book:** Charles Darwin, *The Expression of the Emotions in Man and Animals* (John Murray, 1872).

184 **emotional awareness:** See https://www.paulekman.com for his website. See also https://scholar.google.com/citations?user=ADetnc0AAAAJ&hl=en for his scientific publications in the field of psychology.

184 **chances are others do too:** Robert L. Trivers, *Deceit and Self-Deception: Fooling Yourself the Better to Fool Others* (Allen Lane, 2011).

185 **two constraints on language:** Terrence W. Deacon, *The Symbolic Species: The Co-Evolution of Language and the Brain* (W. W. Norton, 1997).

187 **"in the mind of man":** Charles Darwin, *The Descent of Man, and Selection in Relation to Sex* (John Murray, 1871), 61.

187 **constituent languages:** Afrikaans is no longer the language of the oppressor. Currently more nonwhites than whites speak Afrikaans in South Africa; see https://www.up.ac.za/afrikaans-english/news/post_2465048-more-than-an-oppressors-language-reclaiming-the-hidden-history-of-afrikaans#.

187 **similar set of grammatical rules:** Jared Diamond, *The Rise and Fall of the Third Chimpanzee: How Our Animal Heritage Affects the Way We Live* (Vintage, 2002).

188 **our species emerged:** Daniel E. Lieberman, "Speculations About the Selective Basis for Modern Human Craniofacial Form," *Evolutionary Anthropology: Issues, News, and Reviews* 17, no. 1 (2008): 55–68.

189 **some parts of it:** Alison K. Schug, Edith Brignoni-Pérez, Nasheed Jamal, and Guinevere Eden, "Gray Matter Volume Differences Between Early Bilinguals and Monolinguals: A Study of Children and Adults," *Human Brain Mapping* 43, no. 16 (2022): 4817–34.

189 **larger volume of gray matter:** O. A. Olulade, N. J. Jamal, D. S. Koo, C. A. Perfetti, C. LaSaoo, and G. F. Eden, "Neuroanatomical Evidence in Support of the Bilingual Advantage Theory," *Cerebral Cortex* 26, no. 7 (2015): 3196–204.

190 **changes in the brain:** E. A. Maguire, D. Gadian, I. Johnsrude, C. Good, J. Ashburner, R. Frackowiak, et al., "Navigation-Related Structural Change in the Hippocampi of Taxi Drivers," *Proceedings of the National Academy of Sciences* 97, no. 8 (2000): 4398–403.

190 **Terrence Deacon:** Deacon, *Symbolic Species*.

191 **part of it:** James C. Pang, Kevin Aquino, Marianne Oldehinkel, Peter Robinson, Ben Fulcher, Michael Breakspear, et al., "Geometric Constraints on Human Brain Function," *Nature* 618 (June 2023): 566–74.

SEVEN: OTHER MINDS

196 **before our species evolved:** Daniel L. Everett, *How Language Began: The Story of Humanity's Greatest Invention* (Liveright, 2017).

197 **clinging to debris:** Gerrit van den Bergh, Bo Li, Aam Brumm, Rainer Grün, Dida Yurnaldi, Mark Moore, Iwan Kurniawan, et al., "Earliest Hominin Occupation of Sulawesi, Indonesia," *Nature* 529, no. 7585 (2016): 208–11.

197 **west to east:** Scott A. Hocknull, Philip Piper, Gert van den Bergh, Rokus Due, Michael Morwood, and Iwan Kurniawan, "Dragon's Paradise Lost: Palaeobiogeography, Evolution and Extinction of the Largest-Ever Terrestrial Lizards (Varanidae)," *PLoS ONE* 4, no. 9 (2009): e7241.

198 **religion, arts, and money:** Yuval N. Harari, *Sapiens: A Brief History of Humankind* (Vintage, 2011).

200 **cumulative cultural evolution:** Joseph Henrich, *The Secret of Our Success: How Culture Is Driving Human Evolution, Domesticating Our Species, and Making Us Smart* (Princeton University Press, 2016).

201 **by genetics or natural selection:** Stanley I. Greenspan and Stuart G. Shanker, *The First Idea: How Symbols, Language and Intelligence Evolved from Our Primate Ancestors to Modern Humans* (Da Capo Press, 2004).

203 **"including many members who":** Charles Darwin, *The Descent of Man, and Selection in Relation to Sex* (John Murray, 1871), 167.

203 **caused our peaceful nature:** Richard Wrangham, *The Goodness Paradox: How Evolution Made Us Both More and Less Violent* (Profile Books, 2019).

204 **the help you require:** Richard D. Alexander, *Darwinism and Human Affairs* (University of Washington Press, 1979).

204 **social status within the group:** Nicole H. Hess and Edward H. Hagen, "The Impact of Gossip, Reputation, and Context on Resource Transfers Among aka Hunter-Gatherers, Ngandu Horticulturalists, and MTurkers," *Evolution and Human Behavior* 44, no. 5 (2023): 442–53.

206 **shirking its responsibilities:** Redouan Bshary and Alexandra S. Grutter, "Image Scoring and Cooperation in a Cleaner Fish Mutualism," *Nature* 441 (June 2006): 975–78.

207 **partner will be chosen:** Liran Samuni and Martin Surbeck, "Cooperation Across Social Borders in Bonobos," *Science* 382, no. 6672 (2023): 805–809.

208 **art and jewelery:** Curtis W. Marean, "An Evolutionary Anthropological Perspective on Modern Human Origins," *Annual Review of Anthropology* 44, no. 1 (2015): 533–56.

210 **when language evolved:** Charles Perreault and Sarah Mathew, "Dating the Origin of Language Using Phonemic Diversity," *PLoS ONE* 7, no. 4 (2012): e35289.

212 **necks were too short:** Philip Lieberman and Robert McCarthy, "Tracking the evolution of language and speech," *Expedition*, 49, no. 2 (2007): 15–20.

214 **decay to uracils:** David Gokhman, Malka Nissim-Rafinia, Lily Agranat-Tamir, Genevieve Housman, Raquel García-Pérez, Esther Ligano, et al., "Differential DNA Methylation of Vocal and Facial Anatomy Genes in Modern Humans," *Nature Communications* 11, no. 1 (2020): 1189.

215 **445,000 and 473,000 years ago:** Kay Prüfer, Fernando Racimo, Nick Pattrson, Flora Jay, Sriram Sankararaman, Susanna Sawyer, et al., "The Complete Genome Sequence of a Neanderthal from the Altai Mountains," *Nature* 505, no. 7481 (2014): 43–49.

215 **team of researchers did:** Gokhman et al., "Differential DNA methylation."

216 **the answer is sex:** Geoffrey Miller, T*he Mating Mind: How Sexual Choice Shaped The Evolution of Human Nature* (Vintage, 2001).

218 **a neat experiment:** Jason Keagy, Jean-François Savard, and Gerald Borgia, "Male Satin Bowerbird Problem-Solving Ability Predicts Mating Success," *Animal Behaviour* 78 (October 2009): 809–17.

218 **will mate with the male:** Laura A. Kelley and John A. Endler, "How Do Great Bowerbirds Construct Perspective Illusions?," *Royal Society Open Science* 4, no. 1 (2017): 160661.

219 **twenty-five and a half years:** Data for *Ptilonorhynchus violaceus* (satin bowerbird), Australian Bird and Bat Banding Scheme Database, Australian Government, Department of Climate Change, Energy, the Environment and Water, accessed October 9, 2024, https://www.environment.gov.au/cgi-bin/biodiversity/abbbs/abbbs-search.pl?taxon_id=378.

220 **male's impressive display:** Jürgen Otto, "Peacock Spider 7

(Maratus speciosus)," posted by Peacockspiderman, March 1, 2013, YouTube, https://www.youtube.com/watch?v=d_yYC5r8xMI.

221 **other species of *Homo*:** Prüfer et al., "Complete Genome Sequence."

221 **other human population:** Matthew Silcocks, Ashley Farlow, Azure Hermes, Georgia Tsambos, Hardip Patel, Sharon Huebner, et al., "Indigenous Australian Genomes Show Deep Structure and Rich Novel Variation," *Nature* 624 (December 2023): 593–601.

222 **Luzon, in the Philippines:** Florent Détroit, Armand Mijares, Julien Corny, Guillaume Daver, Clément Zanolli, Eusebio Dizon, et al., "A New Species of Homo from the Late Pleistocene of the Philippines," *Nature* 568, no. 7751 (2019): 181–86.

222 **on his deathbed:** Qiang Ji, Wensheng Wu, Yannan Ji, Qiang Li, and Xijun Ni, "Late Middle Pleistocene Harbin Cranium Represents a New *Homo* Species," *Innovation* 2, no. 3 (2021): 100132.

222 **one headline asked:** Kat J. McAlpine, "Is 'Dragon Man' a Missing Link in Human Evolution?" *The Brink*, Boston University, July 9, 2021, https://www.bu.edu/articles/2021/is-dragon-man-a-missing-link-in-human-evolution/.

EIGHT: BRAVE NEW WORLD

225 **novel *Brave New World*:** Aldous Huxley, *Brave New World* (Chatto & Windus, 1932).

226 **Klara is a humanoid:** Kazuo Ishiguro, *Klara and the Sun* (Knopf, 2021).

227 **1999, a teen died:** Meir Rinde, "The Death of Jesse Gelsinger, 20 Years Later," *Distillations Magazine*, June 4, 2019, https://sciencehistory.org/stories/magazine/the-death-of-jesse-gelsinger-20-years-later/.

229 **facilitates HIV infection:** Vera L. Raposo, "The First Chinese Edited Babies: A Leap of Faith in Science," *JBRA Assisted Reproduction* 23, no. 3 (2019): 197–99.

229 **"three-parent babies":** Edward H. Morrow, Klaus Reinhardt, Jonci N. Wolff, and Damian K. Dowling, "Risks Inherent to Mitochondrial Replacement," *EMBO Reports* 16, no. 5 (2015): 541–44.

NOTES

230 **all highly heritable:** Francis Galton, *Hereditary Genius: An Inquiry into Its Laws and Consequences* (Macmillan, 1869).

230 **a historical perspective:** Carl Zimmer, *She Has Her Mother's Laugh: The Powers, Perversions and Potential of Heredity* (Dutton, 2018).

230 **social structures of ant colonies:** Edward O. Wilson, *Sociobiology: The New Synthesis* (Harvard University Press, 1975).

231 **to fund the project:** Alexander Etkind, "Beyond Eugenics: The Forgotten Scandal of Hybridizing Humans and Apes," *Studies in History and Philosophy of Science Part C: Studies in History and Philosophy of Biological and Biomedical Sciences* 39, no. 2 (2008): 205–21.

235 **our relatives have them, too:** Frans B. M. de Waal, *The Bonobo and the Atheist: In Search of Humanism Among the Primates* (W. W. Norton, 2013).

235 **a sense of fairness:** Sarah F. Brosnan and Frans B. M. de Waal, "Monkeys Reject Unequal Pay," *Nature* 425 (July 2003): 297–99.

236 **video of the experiment:** "Moral Behavior in Animals," TED talk, November 2011, https://www.ted.com/talks/frans_de_waal_moral_behavior_in_animals?language=en.

236 **the long-tailed macaque:** Rowan Titchener, Constance Thiriau, Timo Hüser, HansjOrg Scherberger, Julia Fischer, and Stefanie Keupp, "Social Disappointment and Partner Presence Affect Long-Tailed Macaque Refusal Behaviour in an 'Inequity Aversion' Experiment," *Royal Society Open Science* 10, no. 3 (2023): 221225.

237 **case of a computer:** Daria Knoch, Alverto Pascual-Leone, Kaspar Meyer, Valerie Treyer, and Ernst Fehr, "Diminishing Reciprocal Fairness by Disrupting the Right Prefrontal Cortex," *Science* 314, no. 5800 (2006): 829–32.

238 **trying to lay eggs:** Thibaud Monnin and Francis L. W. Ratnieks, "Policing in Queenless Ponerine Ants," *Behavioral Ecology and Sociobiology* 50, no. 2 (2001): 97–108.

238 **trying to cheat:** Francis L. W. Ratnieks and P. Kirk Visscher, "Worker Policing in the Honeybee," *Nature* 342 (December 1989): 796–97.

238 **above our evolutionary destiny:** Peter Singer, *The Expanding Circle: Ethics and Sociobiology* (Clarendon Press, 1981).

240 **"hungry judge effect":** "Hungry Judge Effect," Wikipedia, n.d., https://en.wikipedia.org/wiki/Hungry_judge_effect#.

240 **Handbook for Judicial Officers:** Judicial Commission of New South Wales, "Judge v. Robot: Artificial Intelligence and Judicial Decision-Making," https://www.judcom.nsw.gov.au/publications/benchbks/judicial_officers/judge_v_robot.html#.

244 **published in 1881:** Charles Darwin, *The Formation of Vegetable Mould Through the Action of Worms, with Observations of their Habits* (John Murray, 1881).

244 **"plasticity first" evolution:** Mary Jane West-Eberhard, *Developmental Plasticity and Evolution* (Oxford University Press, 2003).

245 **encoded into DNA:** Benjamin P. Oldroyd, *Beyond DNA: How Epigenetics Is Transforming Our Understanding of Evolution* (Melbourne University Press, 2023).

245 **a means to speed up evolution:** James Mark Baldwin, "A New Factor in Evolution," *American Naturalist* 30 (1896): 441–51, 536–53.

246 **unconvinced about the Baldwin effect:** Mauro Santos, Eörs Szathmáry, and José Fontanari, "Phenotypic Plasticity, the Baldwin Effect, and the Speeding up of Evolution: The Computational Roots of an Illusion," *Journal of Theoretical Biology* 371 (April 2015): 127–36.

246 **satisfying mechanistic explanation:** Cecilia Heyes, Nick Chater, and Dominic Michael Dwyer, "Sinking In: The Peripheral Baldwinisation of Human Cognition," *Trends in Cognitive Sciences* 24, no. 11 (2020): 884–99.

248 *social curiosity hypothesis*: Sofia Forss and Erik Willems, "The Curious Case of Great Ape Curiosity and How It Is Shaped by Sociality," *Ethology* 128, no. 8 (2022): 552–63.

250 **impossible for the time:** Tony Freeth, "An Ancient Greek Astronomical Calculation Machine Reveals New Secrets," *Scientific American*, January 1, 2022, https://www.scientificamerican.com/article/an-ancient-greek-astronomical-calculation-machine-reveals-new-secrets/.

250 **using their naked eyes:** Tony Freeth, David Higgon, Aris Dacanalis, Lindsay MacDonald, Myrto Georgakopoulou, and Adam

Wojcik, "A Model of the Cosmos in the Ancient Greek Antikythera Mechanism," *Scientific Reports* 11, no. 1 (2021): 5821.

251 **metaphor Peter Singer used:** Peter Singer, *The Expanding Circle: Ethics and Sociobiology* (Clarendon Press, 1981), 88.

252 **and the United States:** Tim McMillan, "New Study Reveals Sharp Decline in American IQ Scores as the 'Reverse Flynn Effect' Takes Center Stage," *The Debrief*, March 29, 2023 https://thedebrief.org/new-study-reveals-sharp-decline-in-american-iq-scores-as-the-reverse-flynn-effect-takes-center-stage/.

253 **have significant effects:** Anders Bergström, Shane McCarthy, Ruoyun Hui, Mohamed Almarri, Qasim Ayub, Petr Danecek, Yuan Chen, et al., "Insights into Human Genetic Variation and Population History from 929 Diverse Genomes," *Science* 367, no. 6484 (2020): eaay5012.

254 **survive these new environments:** Ann Gibbons, "Dinner with Darwin: Sharing the Evidence Bearing on the Origin of Humans," in *A Most Interesting Problem: What Darwin's* Descent of Man *Got Right and Wrong About Evolution*, ed. J. M. DeSilva (Princeton University Press, 2021), 204–22.

254 **interbreeding events with Denisovans:** Emilia Huerta-Sánchez, Xin Jin, Asan, Zhuoma Bianba, Benjamin Peter, Nicolas Vinckenbosch, et al., "Altitude Adaptation in Tibetans Caused by Introgression of Denisovan-like DNA," *Nature* 512, no. 7513 (2014): 194–97.

254 **were particularly at risk:** Hugo Zeberg, "The Major Genetic Risk Factor for Severe COVID-19 Is Associated with Protection Against HIV," *Proceedings of the National Academy of Sciences* 119, no. 9 (2022): e2116435119.

EPILOGUE: MODERN FAMILY

258 **myth of the mother instinct:** Chelsea Conaboy, *Mother Brain: How Neuroscience Is Rewriting the Story of Parenthood* (St. Martin's Press, 2022).

259 **psychologists and sociologists:** Rebecca Sear, "The Male Breadwinner Nuclear Family Is Not the 'Traditional' Human Family, and Promotion of this Myth May Have Adverse Health

Consequences," *Philosophical Transactions of the Royal Society B: Biological Sciences* 376, no. 1827 (2021): 20200020.

260 **become better at fooling us:** Wai Keen Vong, Wentao Wang, A. Emin Orha, and Brenden Lake, "Grounded Language Acquisition Through the Eyes and Ears of a Single Child," *Science* 383, no. 6682 (2024): 504–11.

261 **sounds with objects:** Melis Çetinçelik, Caroline Rowland, and Tineke Snijders, "Do the Eyes Have It? A Systematic Review on the Role of Eye Gaze in Infant Language Development," *Frontiers in Psychology* 11 (January 2021): 589096.

262 **the easier learning becomes:** Patricia K. Kuhl, "Brain Mechanisms in Early Language Acquisition," *Neuron* 67, no. 5 (2010): 713–27.

262 **needed for future learning:** Patrick Butler, "No Grammar Schools, Lots of Play: The Secrets of Europe's Top Education system," *Guardian*, September 20, 2016, https://www.theguardian.com/education/2016/sep/20/grammar-schools-play-europe-top-education-system-finland-daycare.

Index

Pages in *italics* refer to figures.

Africa, 35, 37, 40, 50–51, 59–61, 209–10, 216
Agassiz, Jean Louis Rodolphe, 75–76, 77
AI technology, 233, 239–42
Alexander, Richard D., 202–4, 206
alleles, 91, 107–8
alloparents, 135, 146
ants, 3–4, 6, 57, 58, 125, 238
apes, 53–54, 62, 64, 66, 173–75
AQP7 gene, 118–19
Ardipithecus kadabba, 51
Ardipithecus ramidus (Ardi), 47–48, 50, 51, 53
Aristotle, 25–30, 75, 76
Asia, 32–35, 40–41
assisted-reproductive technologies, 253
associative learning, 104, 162, 163–64
Atkinson, Elizabeth, 168–69
Australopithecus, 34, 51, 55, 56, 59, 158
Australopithecus afarensis (Lucy), 36, 50, 51, 53–56, 59, 139
Australopithecus africanus, 50

babies:
age at weaning, 67–68
brain size of, 65, 66, 104, 200
choosing sex of, 229
facial expressions imitated by, 176, 177–78
head size and, 8, 56
helplessness of, 1, 2, 8, 9, 10, 63–64, 66, 68, 69, 87, 150, 152, 154, 160, 193, 203, 221
language and, 160, 170–73, 175, 180, 181, 185
learning and, 154–55, 160
as mind readers, 2, 178
nursing and, 131
social bonds and, 152, 160, 176–78, 181
See also care of young
Baldwin, James, 245–46
Baldwin effect, 246
bank fraud, 234, 235
Bar mutation, 110–12
Benz, Karl, 242
biofilms, 198
biogenetic law, 77–78

bipedalism:
  balance and stability of, 52–53, 61
  birth canal shape and, 53, 54, 69
  carrying a baby and, 55
  environmental changes and, 139, 157
  foramen magnum and, 74, 83–85
  human evolution and, 50–51, 53, 68, 69, 88, 97, 100, *268–69*
  pelvis and, 63–64
  phenotype plasticity and, 46–47
  running and, 60–61, 62, 64, 100, 118
  segmentation clock and, 81
  skeletal structure and, 70
  social bonds and, 55, 141
  speed sacrificed in, 54, 55
birds:
  brain of, 134–35, 137
  calls and songs of, 133, 134, 170–72
  care of young, 134–35, 143, 145–46
  cues from others and, 247
  FOXP2 gene and, 167
  mating of, 217–19, 220
  signaling pathways of, 126–27
  tools and, 57, 154
Black, Davidson, 33–34
BMP (bone morphogenetic protein) mutation, 85–89
bonobos, 7, 90, 97, 139, 173–75, 206–7, 247–49
bowerbirds, 217–19, 220

brain:
  of babies, 65, 66, 104, 200
  of birds, 134–35, 137
  breathing and, 191–92
  care of young and, 135, 137, 153–54
  cell differentiation within, 105
  of children, 185, 188, 189
  diet and, 62–63, 69
  expansion of, 8, 29, 30, 35–36, 37, 48, 50, 51, 53, 56, 57–63, 100, 104, 114–19, 124, 133–34, 200, 216–17, 221
  as expensive organ, 8, 63, 65, 69–70, 134
  geometry of, 190–91
  gray matter of, 8, 189
  heat exhaustion and, 117–18
  language sculpting brain, 188–92
  of mammals, 113, 116, 153–54
  mental capacity and, 69, 70
  neurons and, 112, 113–15, 188, 190
  sexual selection and mate choice, 216–17
  social brain hypothesis, 136–37
brain collective, 201
breathing, 191–92, 212
breeder's equation, 123–24
Brosnan, Sarah, 235
Bulk, George, 41

capuchin monkeys, 235–36, 237
care of young:
  birds and, 134–35, 143, 145–46
  brain and, 135, 137, 153–54

cold-blooded vertebrates and,
    141–45
grandparents and, 155,
    158–59, 160, 258, 260
human evolution and, 59, 138
language and, 8, 9, 134, 153,
    193, 221, 251–52, 255, 262
mammals and, 1, 143–50
mother instinct and, 257–58,
    259
nuclear family and, 258–60
requirements of, 66, 258
self-domestication and,
    137–38
social bonds and, 70, 152–55,
    159–60, 262
warm-blooded vertebrates and,
    143–44, 145
cassowaries, 126–27, 156, 201
Ceaușescu, Nicolae, 153, 188
Centre for the Study of
    Existential Risk, 233, 239
Charpentier, Emmanuelle, 115
chatbots, 239, 240, 260
cheating, 206, 238, 247
Chiang, Ted, 225
children:
    bipedalism and, 46–47
    brain of, 185, 188, 189
    dependence of, 68
    language and, 170, 175, 185,
        187, 188–89, 199
Child's View for Contrastive
    Learning (CVCL) model,
    260–61
chimpanzees:
    babies of, 63, 64–65
    bipedalism of, 52
    birth canal shape and, 53
    brain of, 116

care of young, 138
chewing of, 120
chromosome pairs of, 90
communication of, 211
cranial base of, 74
curiosity and, 247–49
diet of, 62–63
eye-gazing hypothesis and, 182
facial expressions imitated by,
    176–77
genetic variation among, 100
human ancestors and, 7, 35,
    36, 47–48, 50, 51, 67, 85,
    96–97, 111, 206
human DNA compared to,
    21–25, 29, 42, 43, 215
intelligence of, 57
interbirth interval of, 67
knuckle walking of, 48
language learning and, 163,
    164
mating system of, 97, 147
penile spines of, 86–87
skull of, 121, 127
social brain hypothesis and,
    136–37
tools and, 58
trunk rotation and, 61
Chomsky, Avram Noam, 162–64,
    169–70, 175, 185, 193, 243
chromosome-fusing mutation, 90,
    97–98, 100, *268–69*
chromosomes:
    homologous chromosomes,
        90–91, 109
    number of pairs, 90–99, 109
    structure of, 108–9
cichlids, 142–43
cleaner fish, 204–6
Clutton-Brock, Tim, 149

cognitive abilities, 136–37, 153, 160, 199–200, 202
cold-blood vertebrates, care of young, 141–45
communication:
   animal communication, 162, 169
   ants and, 3–4, 6
   development of, 2, 193
   eye contact and, 181–82
   facial expressions and, 1, 176–78, 180, 183–84
   honeybees and, 4–5, 6, 197
   language's relationship to, 198
   mutual understanding and, 180
   natural selection and, 6, 8, 9
   nonverbal communication, 169, 179–80, 181, 182, 183–84
   positive selection and, 168
   social bonds and, 7
   See also language; speech
Conaboy, Chelsea, 258
cooperation, 202–7
corticobulbar tract, 192
corticospinal tract, 192
COVID-19 symptoms, 254
cranial base, 122–23
cranial base flexion, 123, 126, 128
cranial base inflexion, 74, *75*, 123
cranial vault, 121, 122–23, 126
creationism, 169
creoles, 187
criminally insane patients, 233–34, 235
CRISPR-Cas9, 114–15, 228–29, 231
cultural evolution, 199–202

curiosity, 247–49
cytosines, 214

Dart, Raymond Arthur, 49–50
Darwin, Charles:
   on barnacles, 76–77
   on bipedalism, 68
   on birds, 127
   on common ancestor of vertebrates, 82–83
   on eugenics movement, 230
   on facial expressions, 183
   on families, 13–14
   Forbes Quarry skull and, 41–42
   on HMS *Beagle*, 14, 77
   on human evolution, 25, 26, 29–30, 31, 41–42, 50, 58–59, 76, 183, 243, 251
   Thomas Huxley and, 226
   on inbreeding, 98
   on language, 29–30, 186–87, 207
   on mental capacity, 70
   on natural selection, 14–16, 20, 29–30, 203, 251
   on sexual selection, 219–20
   on shared ancestry, 28
   on social bonds, 69
   on species variation, 14–15
   on worms, 243–44
Dawkins, Richard, 13–16, 24, 222–23
Dawson, Charles, 48–49
Deacon, Terrence, 185, 190
de Waal, Frans, 58, 98, 235–36
diseases, DNA sequencing of, 17–19
DNA:
   chromosome recombination and, 99

codon in, 209
double helix of, 16, 23
epigenetic marks and, 213–14
human DNA compared to chimpanzee DNA, 21–25, 29, 42, 43, 215
junk-DNA, 24
proteins and, 21–24
reading of, 16, 80, 85
sequencing of, 17–19, 22, 23, 24, 28–29, 37–38, 39, 40, 106, 111, 214
Doudna, Jennifer, 115
Dubois, Marie Eugène François Thomas, 30–36, 49–50, 77, 222
dwarfism, 37

Ekman, Paul, 183–84
embryonic development, 73–74, 76–83, 86, 100, 113
empathy, 238, 241
energetics of gestation and growth (EGG) hypothesis, 66
environments:
 adaptation to, 7, 14, 37, 56, 62, 69, 89, 127, 139–41, 157, 200–201, 245, 250–51, 254
 modification of, 253
Eoanthropus dawsoni, 49
epigenetic marks, 213–14, 244
epigenetics, 213–15, 244–45
epiglottis, 131
ethics, of gene therapy, 228–29
Ethiopia, 36, 47–48
eugenics movement, 230, 231
eukaryotic tree, 80
Europe, Neanderthal DNA in, 40

Everett, Daniel, 196–97
evolution:
 of bee communication, 5
 cultural evolution, 199–202
 development of embryos and, 76
 genetics and, 13, 111, 200
 of language, 178, 184, 186–87, 196–97, 251, 255, *268–69*
 manipulation of, 229–31, 232
 neoteny and, 72–73, 76, 105, 121
 spatial packing problem and, 124–28
 See also human evolution; natural selection
eyes:
 eye-gazing hypothesis, 182–83, 261
 sclera of, 181–82

facial muscles, 164, 183, 192
fairness, 235–38
Falconer, Hugh, 41–42
family, changes in, 257–62
feet, shape and form of, 47–48, 53, 88
fire ants, 197
FitzRoy, Robert, 14
Flores island, 35, 37, 196–97
Flynn, James, 252
Flynn effect, 252
foramen magnum, 73, 74, 75, 83–85
founder effect, 210
FOXP2 gene, 166–69, 211
Fraipoint, Julien, 31–32
frogs, 132, 142
fruit flies, 24, 108–11, 112
Fürbringer, Max Carl Anton, 32

gain of function, 88
Galton, Francis, 230, 242
gametes, 91–92, 99, 145, 226
GDF6 (growth differentiation factor 6), 88
gel electrophoresis, 22–23
gene conversion, 112, 119
gene duplication, 106–7, 112, 117, 119
gene expression, 78, 117, 244
gene nursery, 105
gene-regulatory networks, 46–47, 82, 100, 127, 215
genes:
    Richard Dawkins on, 15, 16
    epigenetics and, 213–14
    human genes, 20, 24, 229
    inversion and, 111–12, 115
    methylation of, 214
    modifications to, 228–29, 268–69
    natural selection and, 15, 43, 253–54
    transposons, 110–11
gene sequences, 112
gene therapy, 227–28
genetic drift, 106–7
genetics:
    evolution and, 13, 111, 200, 268–69
    fruit flies and, 109
    of *Homo sapiens*, 9
    honeybees and, 6
    speech and, 164–69, 193, 215
    technical language of, 10
genetic variation, 100
genomes, sequencing of, 16, 18, 19–20, 24, 42–43

genome-wide association studies (GWAS), 17–19
germline, modification of, 228–29
gestation, 65–66, 67, 147
Gokhman, David, 215
Goodall, Jane, 58, 98
Gopnik, Alison, 154
gorillas, 48, 85, 90, 97, 116–17, 138, 156–57
Gould, Stephen Jay, 63–64, 72, 73, 83
Greenspan, Stanley, 201–2
Griesser, Michael, 135

HACNS1 (human-accelerated conserved noncoding sequence 1), 88–89
Haeckel, Ernst, 30–32, 77–78, 81, 201–2
Haile-Selassie, Yohannes, 47
Haldane, John Burdon Sanderson, 203
hands, shape and form of, 53, 58–59, 88
Harari, Yuval Noah, 198–99
Hawking, Stephen, 239
Hawkins, Benjamin Waterhouse, 26
He Jiankui, 228–29
head:
    shape of, 8, 50, 53, 60, 61, 74, 75, 119–24, 126–28, 133, 191
    sweat glands of, 118
heat exhaustion, 117–18
HeLa cell line, 114
hemophilia, 98
Henrich, Joseph, 199–201
heredity monarchy, 105–6

Hes7 gene, 81–82, 113
Hes factors, 113–14
hippocampus, 190
Hobbes, Thomas, 237
Hominidae, 47
hominins, 47–48, 53, 60, 63, 68
Homo:
    bipedalism and, 55, 56, 61–62
    brain size and, 51, 53, 56,
        57–63, 104, 216–17
    crural index ratio and, 56
    cultural evolution and, 200
    diet of, 62, 120–21
    life span and, 158
    pelvic shape and, 55, 56, 61
    running and, 60–61, 62
    transformation of *Homo
        sapiens* and, 198–99,
        268–69
Homo denisova:
    FOXP2 gene and, 168–69, 211
    Homo sapiens interbreeding
        with, 41, 94, 214–15, 216,
        221, 254
    imaginary train journey and,
        196
    skulls of, 212–13
Homo erectus:
    brain size of, 59, 60, 63, 118,
        157
    cooperation and, 141
    diet of, 62, 120
    evolution of, 196–97
    fossil remains of, 34
    Homo florensiensis and, 36–37
    imaginary train journey and,
        196
    language and, 210
    leaving Africa, 35, 59, 60, 216
    running and, 60–61, 62

    skull shape of, 105
    Taung Child compared to, 50
Homo floresiensis (The Hobbit),
    35, 36–37, 196–98, 222
Homo habilis, 56, 59
Homo heidelbergensis, 37, 42,
    104–5, 196
Homo longi (Dragon Man), 222
Homo luzonensis, 222
Homo modjokertensis, 33–34
Homo neanderthalensis:
    DNA of, 40, 41
    FOXP2 gene and, 168–69, 211
    head shape and, 50, 119,
        127–28
    Homo sapiens interbreeding
        with, 39, 94, 214–15, 216,
        254–55
    human evolution and, 32, 37,
        42, 214–15
    imaginary train journey and,
        196
    longevity of, 158
    speech and, 212, 213, 216
Homo sapiens:
    Africa and, 37, 210, 216
    body hair of, 55, 150, 150*n*
    brain size and, 59–60, 62,
        69–70, 104, 119, 125, 157,
        159, 200, 216–17
    as cooperative breeders,
        157–58
    diet of, 62–63, 120–21,
        139–41, 157, 159
    hipbones of, 8, 48, 52–53, 63,
        64, 66, 68–70
    hunting and, 140–41
    interbreeding of, 39, 41, 215,
        216
    Jebel Irhoud fossil, 119, 128

Homo sapiens (cont.)
   language and, 10, 161, 169, 191, 196, 243, 255, *268–69*
   life span of, 158, 160
   Ernst Walter Mayr on, 34
   neurological growth of, 9
   pregnancies of, 64, 65–66
   skeletons adjusted to walking upright, 8, 70
   skull of, 120–23
Homo transvaalensis, 34
honeybees, 4–7, 13–14, 125, 197, 238
Hooker, Joseph Dalton, 42
Hoppius, Christianus Emmanuel, 27
Hox genes, 80–82, 111
Hrdy, Sarah Blaffer, 138, 151
human DNA, chimpanzee DNA compared to, 21–25, 29, 42, 43, 215
human evolution:
   bipedalism and, 50–51, 53, 68, 69, 88, 97, 100
   Charles Darwin on, 25, 26, 29–30, 31, 41–42, 50, 58–59, 76, 183, 243, 251
   future of, 223
   language and, 164–69, 243, *268–69*
   Neanderthals and, 32
   representations of, 25–26, 27, 48
   studies of, 33–42, 47–48
   teeth and, 36, 50–51, 62
Human Genome Project, 16, 19, 20, 24, 167, 231
human origin:
   in Africa, 50–51
   Aristotle's *scala naturae* and, 25–30, 76
   in Asia, 32, 33, 34–35
   changes in skeleton and, 7, 8, 37, 70, 100
   malleable pelvis and, 52–56, 63–64
   tools and, 36, 49, 51, 57, 139–40
   tree dwellers and, 7, 9, 47–48, 51, 53, 54–55, 61, 69, 88, 97, 100, 157, 245
   walking upright and, 7, 8, 35, 36
hunter-gatherer societies, 66–67, 140, 154
Huxley, Aldous, 225–26
Huxley, Thomas H., 26–27, 30–31, 226

incest, 98–99
incipient species, 95
industrialization, 258
IQ scores, 252
irrational behavior, 102–4
Ishiguro, Kazuo, 226–27
Ivanov, Illia, 231–32

Japanese pufferfish, 144–45, 147, 220
Java, 32–33

King, Mary-Claire, 21–24, 42
Klinefelter syndrome, 94
knapsack problem, 125
KNM-ER 2596 fossil, 159
koalas, 155, 156
Kubrick, Stanley, 232–33

Lacks, Henrietta, 114, 114*n*
lactase gene, 253–54
Lamarck, Jean-Baptiste, 46

language:
  accumulated knowledge and, 250
  associative learning and, 162, 163–64
  babies and, 160, 170–73, 175, 180, 181, 185
  body language, 180, 183–84
  brain sculpted by, 188–92
  care of young and, 8, 9, 134, 153, 193, 221, 251–52, 255, 262
  children and, 175, 185, 187, 188–89, 199
  collective belief and, 199, 243
  comprehension of, 174–75
  Charles Darwin on, 29–30, 186–87, 207
  evolution of, 178, 184, 186–87, 188, 196–97, 251, 255, *268–69*
  function of, 162, 243
  gossip and, 202, 204, 207, 243
  grammar and, 162, 165–66, 175, 187
  Homo sapiens and, 10, 161, 169, 196, 243, 255
  learning as adult, 180–81, 185–86
  morality and, 238
  natural selection and, 8, 9, 170, 193, 243, 251
  origin of, 161–63, 169, 207–16, 246
  phonemes and, 209–10, 211
  proto-language, 161
  social bonds and, 152, 181, 199, 243
  social reinvention of, 201–2
  storytelling and, 198–99, 202, 243

  symbolic thinking and, 208
  syntax and, 133–34, 184, 198
  viral transmission of, 184–87, 188
  See also communication; speech
language-acquisition device, 162, 163, 170, 185
larynx, 8, 29, 31–32, 131–32, 211–13, *268–69*
Lathrop, Abbie E. C., 85
Lieberman, Daniel, 118
Linnaeus, Carl, 26–28, 33, 220
Lohest, Max, 31–32
Lorenz, Konrad, 151–52

McClintock, Barbara, 111
macrocephaly, 115
mammals:
  aquatic mammals, 192
  brain of, 113, 116, 153–54
  care of young, 1, 143–50
  cuteness of young, 150–58
  gestation period of, 65–66, 67, 147
  head shape of, 60
  Hes7 gene and, 81
  penile spines of, 86–87
  social bonds of, 147–50
  weaning of young, 68
Martínez-Abadías, Neus, 122
matrilineal most recent common ancestor (MRCA), 39
Matsuzawa, Tetsuro, 176–77
Mayr, Ernst Walter, 34, 36
medulla oblongata, 192
meerkats, 148–50, 238
meiosis, 91
menopause, 160
methylation, 214–16

microcephaly, 115
Miller, Geoffrey, 216–17, 219–21
missing link, search for, 30–34, 48–49, 222–23
Mitochondrial Eve, 39–40
mitochondrial sequences, 39–40
monogamous pair bonds, 87
morality:
   AI technology and, 239–40, 242
   criminally insane patients and, 233–34, 235
   empathy and, 238, 241
   fairness and, 235–38
Myowa, Masako, 177

natural selection:
   brain size and, 136
   communication and, 6, 8, 9
   Darwin on, 14–16, 20, 29–30, 203, 251
   developmental shifts and, 126, 193
   FOXP2 gene and, 169
   gene duplication and, 107–8
   gene expression and, 78
   genes and, 15, 43, 253–54
   junk-DNA and, 24
   language and, 8, 9, 170, 193, 243, 251
   social bonds and, 177
   Wallace on, 29
neoteny, 72–73, 76, 105, 121, 268–69
neural plasticity, 189–90
neuroepithelial cells, 116–17
neurogenesis, 112
neurons, 112, 113–15, 188, 190
neuroplasticity, 160

New Caledonian crows, 57–58, 135, 154
Newton, Isaac, 233
NOTCH2 gene, 112, 114–15
NOTCH2NL gene, 115, 117, 119
Notch gene family, 112, 113, 114, 268–69
Notch signaling pathway, 82, 112, 113
nuclear genome sequences, 40
nucleotides, 16–17, 23, 80

obstetrical dilemma, 64, 66
Oldroyd, Ben, 5–6, 102, 148, 186, 212, 245
optimization algorithms, 124–25
orangutans, 85, 138, 156–57, 247–49
Orrorin tugenensis, 51

Pääbo, Svante, 38–41
Paine, Thomas, 105
paleogenetics, 38–39
pandas, 155–56
parrots, 133, 135
Pauling, Linus, 39
Pedals (black bear), 45–47, 74–75
Petrovic, Branko, 192
pharynx, 211
phenotypes, 110, 124, 214
phenotypic plasticity, 46, 244–45
pheromones, 3–4
phonemes, 209–10, 211
Piltdown Man, 48–49
Pinker, Steven, 170, 174
Pithecanthropus, 31
Pithecanthropus erectus (Java Man), 33, 50

plants, phenotypic plasticity of, 46
Plato, 250
Plomin, Robert, 16, 18
polyploid cells, 109
positive selection, 168
proprioception, 178
proteins, 21–24, 80
Putnam, Hilary, 43

quadrupedalism, 84

rabbits, 22, 107–8
radial glia cells, 112, 113–16
recapitulation, 76, 77
reciprocal altruism, 136–37
Reich, David, 40
relativity bias, 101
Romanian orphanages, 152–53, 188

Sagan, Carl, 198
Sahelanthropus tchadensis, 51, 74
Sambourne, (Edward) Linley, 243
SARS-CoV-2, 254–55
Savage-Rumbaugh, Sue, 173–75, 176
Schreiweis, Christiane, 167
segmentation clock, 81
self-domestication, 137–39
selfish gene, 42–43
self-sacrificial behavior, 13–14
serum albumin, 22
sex cells, 90–91
sex ratio, 229
sexual dimorphism, 220
sexual reproduction, 90–91
sexual selection, 216–17, 219–21
Shakespeare, William, 183
Shanker, Stuart, 201–2
SHH gene, 79–80

signaling molecules, 85–86
signaling pathways, 82, 112, 113, 126–27
Sinanthropus lantianensis (Lantian Man), 33
Sinanthropus nankinensis (Nanjin Man), 33
Sinanthropus pekinensis (Peking Man), 33
Sinanthropus yuanmouensis (Yuanmou Man), 33
Singer, Peter, 238, 251
single nucleotide polymorphism (SNP), 17–19, 21
skeletal structure, adaptation of, 7, 8, 37, 45–46, 47, 70, 100
Skinner, B. F., 162–64
slime molds, 101–4
social bonds:
    babies and, 152, 160, 176–78, 181
    bipedalism and, 55, 141
    birth canal shape and, 54, 69
    building of, 2, 7
    care of young and, 70, 152–55, 159–60, 262
    cooperation among non-family members, 202–7
    language and, 152, 181, 199, 243
    longevity and, 158–59
    of mammals, 147–50
    mental capacities and, 69
    natural selection and, 177
    reputation building and, 204–6, 207
    social brain hypothesis, 136–37, 262
    social curiosity hypothesis, 248–49

social disappointment, 236
social engineering, 230
social insects, 4–5, 6, 7, 13–14, 124, 197, 238
spatial packing problem, 124–28
spatial sorting, 96, 97
speech:
   breathing and, 191–92, 212
   comprehension and, 175
   cultural evolution and, 202
   defining of, 131–32
   effect on brain, 188, 191
   fossil evidence of, 10, 207, 211
   genetics and, 164–69, 193, 215
   larynx and, 132, 191, 211–13
   neck length and, 211–13
   tongue and, 132–33, 211, 213
   vocal cavity size and, 211–12
   See also communication; language
SPOCH1 gene, 165–66
stem cells, 112, 113
Stokes, Pringle, 14
Sturtevant, Alfred Henry, 108–10, 112

Taung Child, 49–50
teeth, 36, 50–51, 62, 68, 158
theory of mind, 2, 179–80
thermoregulation, 120
Tinbergen, Niko, 13, 151
Tolkien, J. R. R., 3, 35
Tomasello, Michael, 179–80
tools, 36, 49, 51, 57–58, 139–40, 154, 207–8

totipotent cells, 113
trade, 208
transhumanism, 240–41
Trivers, Robert, 184
trust, 136–39, 172, 177, 184, 208
Turner syndrome, 94

Ultimatum Game, 236–37
unequal crossing-over, 110, 111

Victoria, Queen of England, 98–99
Virchow, Rudolf, 31–32
Vivaldi, Antonio, 191
vocal tract, 132–33, 211–12
Vogt, Karl, 30
von Frisch, Karl, 4–5, 151
von Koenigswald, Gustav Heinrich Ralph, 33–34

Wallace, Alfred Russel, 29, 161–62, 164, 169, 175, 193
Wedgwood, Emma, 98
West-Eberhard, Mary Jane, 244–45
Wilson, Allan, 21–24, 38, 40, 42
Wilson, Edward O., 230–31
worms, 243–45
Wrangham, Richard, 203–4

Y-chromosomal Adam, 40

ZEB2 gene, 116–17
Zimmer, Carl, 230
Zuckerkandl, Emile, 39

About the Author

Madeleine Beekman is professor emerita of evolutionary biology and behavioral ecology at the University of Sydney, Australia. In the academic year of 2020–21, Beekman was a resident fellow at the Institute for Advanced Study in Berlin (Wissenschaftskolleg zu Berlin), which is where the idea for this book was born. She currently lives with her husband in Australia's northern tropical rainforest, where she can observe the endangered cassowary from her office. She has two adult daughters and a tiny granddaughter.